Darwinizing Culture

Darwinizing Culture: The Status of Memetics as a Science

Edited by

ROBERT AUNGER

Biological Anthropology,
University of Cambridge

With a Foreword by
Daniel Dennett

OXFORD

UNIVERSITY PRESS

OXFORD
UNIVERSITY PRESS

Great Clarendon Street, Oxford OX2 6DP

Oxford University Press is a department of the University of Oxford.
If furthers the University's objective of excellence in research, scholarship,
and education by publishing worldwide in

Oxford New York

Athens Auckland Bangkok Bogotá Buenos Aires Calcutta
Cape Town Chennai Dar es Salaam Delhi Florence Hong Kong Istanbul
Karachi Kuala Lumpur Madrid Melbourne Mexico City Mumbai
Nairobi Paris São Paulo Singapore Taipei Tokyo Toronto Warsaw

with associated companies in Berlin Ibadan

Oxford is a registered trade mark of Oxford University Press
in the UK and in certain other countries

Published in the United States
by Oxford University Press, Inc., New York

A catalogue record for this title is available from the British Library

Library of Congress Cataloging in Publication data
Darwinizing culture: the status of memetics as a science/edited by Robert Aunger.
Includes bibliographical references and index.
1. Social perception. 2. Memetics. I. Aunger, Robert.
HM1041 .D37 2001   302'.12–dc21   00-055729

1 3 5 7 9 10 8 6 4 2

ISBN 0 19 263244 2

Typeset by
Florence Production Ltd, Stoodleigh, Devon

Printed in Great Britain on acid free paper by Biddles Ltd., Guildford & King's Lynn

Contents

Foreword

Daniel Dennett

If there is one proposition that would-be memeticists agree on, it is that the flourishing of an idea – its success at replicating through a population of minds – and the value of an idea – its truth, its scientific or political or ethical excellence – are only contingently and imperfectly related. Good ideas can go extinct and bad ideas can infect whole societies. The future prospects of the meme idea are uncertain on both counts, and the point of this book is not to ensure that the meme meme flourishes, but to ensure that *if it does, it ought to*. It works towards this worthy end by creating a landmark, a fixed point not of doctrine but of evidence and methods, some shared acknowledgment among some leading proponents and critics about how the issues ought to be addressed.

The annual Superbowl of American football draws a huge television audience, and as a result attracts advertisers who are willing to pay more than a million dollars for half a minute of the viewers' distracted attention. In the last few years, an interesting subspecies of Superbowl advertisers has sprung up: the fledging 'dot.com' Internet companies that pour a substantial portion of their initial capitalization into a single make-or-break Superbowl debut, hoping that this brief exposure will launch them safely into the competitive future. Why don't they just advertise on the Internet, their chosen field of battle? A similar question was raised a few years earlier about *Wired*, the (traditional, printed-on-paper, on-sale-at-newsstands) magazine of the Internet. What do these traditional media offer that is not (yet) available on the Internet? For one thing, they offer the guarantee of *shared attention*. When you watch an ad during the Superbowl, you know that you are seeing the same ad, at the same time, as millions of other viewers, and you know that they know this as well. When you see stacks of the same magazine at every newsstand, you know that when you read it that you are not alone in reading it; many, many others will read or have already read the very sentence you are reading. These evanescent communities of shared – and knowingly shared – attention play a crucial role in engendering hard-to-achieve confidence in the message, however trivial the topic. They do this by promising a plethora of paths for coordinating distributed intelligence, making it possible for people to compare notes, pool their knowledge, confirm or disconfirm their individual opinions. It's not that people recognize this promise and reflect on it – and of course they almost never act on it, pursuing those paths of enquiry – but they just somehow feel better, knowing that they are part of a large audience, and this is why they are in fact right to feel better: it is hard to get away with telling a lie in such a public arena. If you stumble upon a tempting but improbable claim during the Superbowl program, you may be skeptical, but at least you will realize (probable subliminally, without articulating it) that the advertiser has risked a contagion of disbelief by broadcasting instead of narrowcasting, this message. A website may reach five million people, but they all engage, in effect, in five million private communications. We may all be getting the same message, but unless we know this, we won't reap the benefits of truly shared intelligence. As the idiom goes, it helps to know that we are all on the same page.

The advertising that goes on everywhere in science – all those vigorous campaigns mounted on behalf of theories or hypotheses – avoid degenerating into mere propagandizing because the academy creates structured networks of knowingly shared attention and mutual knowledge, so that more or less everybody can be on the same page. It is not enough that a thousand clever thinkers have read many of the same books and articles and come to similar conclusions about them; they must know that this is so. There needs to be a scientific community.

Within such a community controversy can reign without rancor and constructive disagreement can prosper, because approximately all the accumulated knowledge of the participants can be brought to bear on a few focal points, a *competitive* but also *concerted* effort. Now that there are more than a handful of serious contenders in the form of partisan proposals (see the bibliographies of the chapters), it is time to start sorting them out. A start is all. I am not entirely persuaded by *any* of the chapters in this book, but this Foreword is not the time and place for me to take issue with them. This Foreword is the time and place for me to applaud the fact that serious consideration of the meme meme is now underway at last, after several decades of relatively ineffectual campaigns by proponents and critics. The workshop from which this volume springs was heated but constructive, and now a wider audience can get on the same page. It will be the first of many, I predict.

Skeptics may be tempted to think that my Foreword itself demostrates the futility of the idea of memetics, by emphasizing the underlying rationality, the intentionality, of the purported 'vectors' of the meme meme. How can mindless Darwinian algorithms cope with such mindful culture-makers, subliminally sensitive to such issues as whether or not the environment includes many paths for coordinating distributed intelligence? But in fact, evolutionary approaches to such underlying conditions of rationality have been leading the way, illuminating the background conditions for communication, cooperation, the establishment of norms and customs, and other phenomena familiar to students of culture. The open question is not whether there will be a Darwinian theory of culture but what shape such a Darwinian theory will take.

It is obvious that there are patterns of cultural change – evolution in the neutral sense – and any theory of cultural change worth more than a moment's consideration will have to be Darwinian in the minimal sense of being *consistent with* the theory of evolution by natural selection of *Homo sapiens.* The demand of this minimal Darwinism are far from trivial, and the ferocity with which Darwinian accounts of the evolution of language and sociality are attacked by some critics from the humanities and social sciences show that in some influential quarters, mere consistency with evolutionary theory is not yet the accepted constraint it ought to be. This is a fact of life that we must deal with: fear of the thin edge of the wedge misleads many who hate the idea of a *strong* Darwinian theory of cultural evolution to resist conceding even consistency with evolutionary theory as the obvious requirement it is. In this volume, minimal Darwinism is taken for granted; no skyhooks are sought within its pages. But there are still plenty of grounds on offer in criticism of various versions of the strong Darwinian thesis of memetics. It will be most interesting to see what settles out of this new exploration.

August 2000

List of Contributors

Robert Aunger was, until recently, Senior Fellow in Cognition and Evolution at King's College, Cambridge. He is now affiliated with the Department of Biological Anthropology there. His PhD, in anthropology, is from the University of California at Los Angeles. His research focuses on the empirical study of cultural transmission, developing reliable methods for ethnography, and theories of cultural evolution. He has taught at Northwestern University, the University of Chicago, as well as Cambridge, and has several books in the works.

Susan Blackmore is Reader in Psychology at the University of the West of England, Bristol, where she teaches parapsychology and consciousness. She has a degree in psychology and physiology from Oxford, an MSc from Surrey University, and gained one of the first PhDs in parapsychology in the country. Her research interests include altered states of consciousness, the effects of meditation, evolutionary psychology, and the theory of memetics. She is author of more than fifty scientific articles. Her books include *Beyond the Body* (1982), *Dying to Live: Science and the near-death experience* (1993), (with Adam Hart-Davis) *Test your Psychic Powers* (1995), and an autobiography *In Search of the Light* (1996). She has been training in Zen for many years. She writes for several magazines and newspapers, and is a frequent contributor and presenter on radio and television. Her most recent book, *The Meme Machine*, was published by Oxford University Press in 1999.

Maurice Bloch is Professor of Anthropology at the London School of Economics and Political Science, as well as a Fellow of the British Academy. He has written many books and articles. Recently, he has been concerned with the relation of anthropology and cognitive psychology. His most recent book is *How We Think They Think: Anthropological approaches to Cognition, Memory and Literacy* (Westview Press, 1999).

Robert Boyd received his bachelor's degree in physics from the University of California at San Diego and a PhD in ecology at UC Davis. He has taught at Duke and Emory Universities and has been at UCLA since 1986. His research focuses on population models of culture summarized in his book, co-authored with P. J. Richerson, *Culture and the Evolutionary Process*. He has also co-authored an introductory textbook in biological anthropology, *How Humans Evolved*, with his wife, Joan Silk.

Rosaria Conte is head of the Division 'AI, Cognitive and Interaction Modelling' at the Institute of Psychology of the Italian National Research Council, and teaches social psychology at the University of Siena. She is quite active in the fields of Agent and Multi-Agent Systems, and Social Simulation. Her research interests range from modelling intelligent agents in interaction to the emergence and evolution of social institutions. She has edited several books and co-authored (with Cristiano Castelfranchi) *Cognitive and Social Action.*

David L. Hull received his PhD in 1964 from the Department of History and Philosophy of Science at Indiana University. His was the first doctorate awarded by this newly formed department. He taught at the University of Wisconsin at Milwaukee from 1964 until he joined the Department of Philosophy at Northwestern University in 1985. He has published a dozen books and anthologies, a hundred papers, and over a hundred book reviews. He has edited forty books in his *Series on the Conceptual Foundations of Science* at the University of Chicago Press. He has been president of the Society of Systematic Zoology, the Philosophy of Science Association, and the International Society for the History, Philosophy and Social Studies of Biology. He has published primarily in biological systematics, evolutionary biology, and philosophy of biology. Recently, he has been working on the nature of science as a selection process.

Adam Kuper is Professor of Social Anthropology at Brunel University. He has done field research in the Kalahari and in Jamaica, and taught in universities in the United States, Sweden, the Netherlands, South Africa, and Uganda, as well as in the United Kingdom. He is the author of a number of books on the history and theory of anthropology, most recently *Culture: The Anthropologists' Account* (Harvard University Press, 1999).

Kevin Laland is a Royal Society University Research Fellow at the Sub-Department of Animal Behaviour, University of Cambridge, where he studies animal behaviour and evolution. He was educated in Psychology at the University of Southampton (BSc) and University College London (PhD). This was followed by a Human Frontier Science Program Fellowship held at the Biology Department of the University of California, Berkeley, and then a BBSRC fellowship in the Zoology Department at Cambridge. He is the author of a substantial number of empirical and theoretical articles on social learning, cultural evolution, and niche construction.

John Odling-Smee studied psychology at University College London, where he worked extensively on animal learning and the role of learning in evolution. He published empirical and theoretical articles, the latter frequently in collaboration with Henry Plotkin. Later, his work expanded to incorporate niche construction, and led to an initial article on that subject in *The Role of Behavior in Evolution* (MIT Press, 1998). More recently, a Leverhulme fellowship led to his present collaboration with Kevin Laland and Marc Feldman, and to further work on niche construction, including a recent paper in the journal *Behavioral and Brain Sciences* (Vol. 23, 2000), which focuses on how niche construction can influence gene–culture coevolution in humans. He currently teaches human sciences students about human evolution at Oxford University.

Henry Plotkin is currently Professor of Psychobiology at University College London and Scientific Director of the ESRC Research Centre on Economic Learning and Social Evolution. He graduated with a First degree in Zoology and Psychology from a South African university. His PhD research was in physiological psychology at the University of London. He has worked on a range of different species, including planaria (a flat-worm), a species of predacious beetle, and various mammalian species including monkeys and humans. He has been working in evolutionary psychology for nearly thirty years and written two books in the area: *Darwin Machines and the Nature of Knowledge*, and *Evolution in Mind*. He is currently writing a book on merging the social and biological sciences.

Peter J. Richerson received a BS (Entomology, 1965) and PhD (Zoology/Limnology 1969), both from the University of California at Davis, and became a member of the faculty of the Department of Environmental Science and Policy at Davis in 1971. He first discovered the need for a real theory of cultural evolution when teaching his first course on principles of human ecology. He and Robert Boyd began working on dual inheritance models shortly thereafter, leading to their first papers in the mid 1970s and eventually to their book *Culture and the Evolutionary Process*. Currently, his main interest is in applying dual inheritance theory to an understanding of the major features of human evolution such as the origin of agriculture and of complex societies. He continues to practice a little limnology.

Dan Sperber is a French social and cognitive scientist. He is the author of *Rethinking Symbolism* (1975), *On Anthropological Knowledge* (1985), *Explaining Culture* (1996), and, with Deirdre Wilson, of *Relevance: Communication and Cognition* (1986, second revised edition, 1995). He holds a research professorship at the Centre National de la Recherche Scientifique (CNRS) in Paris, and has held visiting positions in anthropology, law, linguistics, philosophy, and psychology at Cambridge University, the British Academy, the London School of Economics, the Van Leer Institute in Jerusalem, the Institute for Advanced Study in Princeton, Princeton University, the University of Michigan, and the University of Hong Kong.

Acknowledgements

A number of people and institutions helped to bring about this volume. Professor Ian Donaldson, Convener of the King's College Research Centre, as well as the other Managers of the Research Centre, sponsored the conference on which this book is based, while the British Academy provided additional financial support for the proceedings. Robert Foley generously allowed the King's College financing to be charged to his project on Human Diversity at the Research Centre. Dan Dennett and Kevin Laland provided welcome assistance with the organization of the conference. A variety of anonymous reviewers of the book helped to improve it. Thanks are due to the contributors, not only for their memes, but for being professional and prompt in their delivery of them. Thanks also to Martin Baum at Oxford University Press for his enthusiasm about memes. Finally, I would like to thank the Fellowship of King's College for their tolerance and support of somewhat controversial intellectual interests, evidenced by their admitting me into their Fellowship.

CHAPTER 1

Introduction

Robert Aunger

A number of prominent academics have recently argued that we are entering a period in which evolutionary theory is being applied to every conceivable domain of inquiry. Witness the development of fields such as evolutionary ecology (Krebs and Davies 1997), evolutionary economics (Nelson and Winter 1982), evolutionary psychology (Barkow *et al.* 1992), evolutionary linguistics (Pinker 1994), literary theory (Carroll 1995), evolutionary epistemology (Callebaut and Pinxten 1987), evolutionary computational science (Koza 1992), evolutionary medicine (Nesse and Williams 1994) and psychiatry (McGuire and Troisi 1998)—even evolutionary chemistry (Wilson and Czarnik 1997) and evolutionary physics (Smolin 1997). Such developments certainly suggest that Darwin's legacy continues to grow. The new millennium can therefore be called the Age of Universal Darwinism (Dennett 1995; Cziko 1995).

What unifies these approaches? Dan Dennett (1995) has argued that Darwin's 'dangerous idea' is an abstract algorithm, often called the 'replicator dynamic'. This dynamic consists of repeated iterations of selection from among randomly mutating replicators. Replicators, in turn, are units of information with the ability to reproduce themselves using resources from some material substrate. Couched in these terms, the evolutionary process is obviously quite general. For example, the replicator dynamic, when played out on biological material, such as DNA, is called natural selection. But Dennett suggests there are essentially no limits to the phenomena that can be treated using this algorithm, although there will be variation in the degree to which such treatment leads to productive insights.

The primary hold-out from 'evolutionarization', it seems, is the social sciences. Twenty-five years have now passed since the biologist Richard

Dawkins introduced the notion of a meme, or an idea that becomes commonly shared through social transmission, into the scholastic lexicon. However, the lack of subsequent development of the meme concept has been conspicuous. This stagnation implies that memetics is a what the philosopher Imre Lakatos (1970) would call a 'non-progressive research program'. In particular, there has been no extensive intellectual campaign to produce a general theory of cultural replicators. As will become evident later in this book, little enthusiasm for the meme concept can be found among those professionally charged with understanding culture: that is, cultural and social anthropologists. Those in the fine arts are quite hostile as well. Jaron Lanier (1999), the inventor of the term 'virtual reality,' has argued that 'the notion is so variable as to provide no fixed target . . . Are memes a rhetorical technique, a metaphor, a theory, or some other device? Depending on who you talk to, they can be so wispy as to be almost nothing . . . They make no predictions and cannot be falsified. They are no more than a perspective'. Similarly, the famous skeptic Martin Gardner (2000) recently averred that 'memetics is no more than a cumbersome terminology for saying what everybody knows and that can be more usefully said in the dull terminology of information transfer . . . A meme is so broadly defined by its proponents as to be a useless concept, creating more confusion than light, and I predict that the concept will soon be forgotten as a curious linguistic quirk of little value'. In this view, the analogy to genes is deceptive, and the meme concept is Dawkins' dangerous idea.

At the same time, there are others at the opposite end of the spectrum who herald memes as the saviors of the social sciences. They tout memes as the explanation not only for culture, but for consciousness and the self (e.g., Blackmore 1999). A cottage industry has grown up around the meme idea, with an electronic journal (the *Journal of Memetics–Evolutionary Models of Information Transmission*) and accompanying bulletin board, as well as more standard, printed fare (e.g., Brodie 1996; Lynch 1996; Westoby 1996). Memetics is certainly alive on the World Wide Web and in the popular bookstores, and has considerable currency in some circles, especially among computer literati. This suggests a progressive research program at work.

This image is somewhat illusory, however, as most of the existing work in memetics remains largely abstract. Even those ostensibly sympathetic to the memetic project have noted that there are problems with memes, when considered the focus of an evolutionary process. Dawkins himself has suggested that the meme–gene analogy 'can be taken too far if we are not careful' (Dawkins 1987: 196). Thus, many of the prominent figures in memetics discount the likelihood of memetics ever maturing into an overarching science of culture. They contend that the memetic perspective has yet to enhance our understanding of social-psychological-cultural phenomena compared to more standard formulations such as functionalist or structuralist anthropology. Memetics is surely a very immature science at present, if a science at all.

So what are the specific problems these knowledgeable critics identify? Although a prominent proponent of the memetic perspective, Dennett (1995) has nevertheless mounted perhaps the best-developed attack on the idea that memetics can ever become a science. He primarily elaborates points made earlier by Dawkins himself (e.g., see Dawkins 1982). Most fundamentally, he argues that 'what is preserved and transmitted in cultural evolution is *information*—in a media-neutral, language-neutral sense. Thus the meme is primarily a *semantic* classification, not a *syntactic* classification that might be directly observable in "brain language" or natural language' (Dennett 1995: 353–4; emphasis in original). The syntactic language of genes is in the vocabulary of DNA; that of computer viruses in the computer language that codes it. But if memes exist in the brain, we are unlikely to ever be able to read out the memetic content of some section of the cortex. This suggests to Dennett that social scientists will never have the 'reductionistic' techniques available that biological and physical scientists have used to such effect in finding just how genes replicate using the material substrate of DNA. And even if we find such a technique, we will still need a translation table to convert into a common system of meaning the various media in which the same meme might be represented (in a mind, in splotches of ink on a page, or the digital bits of a computer hard disk).

Dennett then argues that in various ways memes fail to count as proper replicators. First, replicators need high fidelity replication. Memes, however, are subject to high rates of mutation, precluding the establishment of long-lived cultural traditions. Second, these mutations may be directed by purposeful human decision-making among competing cultural alternatives, rather than being simply random choices as expected by Darwinian theory. This is one of the interpretations of what Lamarckism means, with all of its negative connotations (Dennett 1995: 355).

Third, when memes get together in the mind, they mix and match, serendipitously, to fit circumstances, or even accidentally. They do not remain independent particles. Dennett (1995: 355) cites Stephen J. Gould as saying: 'The basic topologies of biological and cultural change are completely different. Biological evolution is a system of constant divergence without subsequent joining of branches. Lineages, once distinct, are separate forever. In human history, transmission across lineages is, perhaps, the major source of cultural change'. So, where biological evolution is slow enough for adaptations to accumulate, and for the selective factors to be identified and ecological correlations noted, evolution in memes is too fast and too combinatorial for selective pressures to have a consistent effect (Dennett 1995: 356).

Fourth, all this rambunctiousness means that similar memes will often crop up, but not be related—rather, they will be invented by clever human brains in similar circumstances by convergent evolution. But we have no good way to determine which memes share ancestry since the tracks they leave behind are mired by replication in different media (Dennett 1995: 356). In conclusion, 'even if memes *do* originate by a process of "descent with modification", our chances of cranking out a science that charts that descent are slim' (Dennett 1995: 356).

However, all of Dennett's arguments constitute empirical claims about aspects of meme transmission and replication parameters which may or may not be true. Little attention has actually been paid to establishing the validity of these assertions, seemingly because they are intuitively obvious. But this does not mean they should be immune from testing. Dennett's claims may only indicate that there are a lot of poorly functioning memes out there; they do not invalidate the meme concept, or prove the impossibility of 'good memes' (Lake 1999).

The question I suggest the thoughtful reader should keep in mind is therefore: Whither memetics? The task of this volume is to see where a reasonable consensus might fall on this spectrum of opinion regarding the utility of the meme concept. As might be expected, perhaps the most interesting terrain lies squarely in the middle—in the temperate zone between the extremes of hot and cold. And, as noted above, some of the middle-ground is taken (in more critical moments) by those who are memes' most ardent defenders.

Perhaps most important in the future development of memetics will be to determine its proper direction. What should be the ambition of memetics? If it is to become a successful science, what is its rightful domain—does it cannibalize the social and psychological sciences *in toto* (as some argue), or should it seek to digest some smaller corner of those provinces, such as social psychology?

What is a meme?

Determining whether memes can account for a relatively wide range of phenomena vitally depends on defining what memes are. Richard Dawkins (1982: 109) suggests a meme is 'a unit of cultural inheritance ... naturally selected by virtue of its "phenotypic" consequences on its own survival and replication' or 'a unit of information residing in a brain'. A more formal definition along this line has been put forward by Aaron Lynch (1998):

MEME: A memory item, or portion of an organism's neurally-stored information, identified using the abstraction system of the observer, whose instantiation depended critically on causation by prior instantiation of the same memory item in one or more other organisms' nervous systems.

The by-now classic examples of memes, according to Dawkins (1976: 206), are 'tunes, ideas, catch-phrases, clothes fashions, ways of making pots or of building arches'. Dawkins (1976: 206) also suggested that memes 'propagate themselves in the meme pool by leaping from brain to brain via a process which, in the broad sense, can be called imitation'. This orthodoxy has been upheld by arguably the most significant English-language works in recent memetics, Dennett's (1995) *Darwin's Dangerous Idea* and Susan Blackmore's *The Meme Machine* (1999).

However, these canonical statements regarding the nature of memes and their mechanism of replication have been contested by others in the field. For example, Gatherer (1998) takes a behaviorist, rather than mentalist, stance toward memes. He takes his inspiration from Benzon (1996: 323):

I suggest that we regard the whole of physical culture as. ... [memes]: the pots and knives, the looms and cured hides, the utterances and written words, the ploughshares and transistors, the songs and painted images, the tents and stone fortifications, the dances and sculpted figures, all of it. For these are the things which people exchange with one another, through which they interact with one another. They can be counted and classified and variously studied.

Memes, in this view, are a heterogeneous class of entities, primarily including behaviors and artefacts—the observable things that permit empirical work. But 'outside the occurrence of the event, the practice of the behaviour, or the lifetime of the artefact, the meme has no existence. The meme does not "go anywhere" when it is not manifested. It is not stored in some neural data bank, some internal meme repository' (Gatherer 1998). Gatherer adopts this stance, largely instrumentally (Gatherer 1999), because neuroscience suggests it is highly unlikely there are replicating information structures in brains (a point seconded by Dennett 1995). In Gatherer's view, the behaviorist position has a number of appealing qualities compared to mentalism, which requires that unobservables (mental states) be taken as the fundamental units of analysis, leading to the empirical doldrums currently experienced by memetics. Since memetics is a cultural, not psychological science, it should aim in his view to describe change in populations by counting up cultural phenomena like artefactual forms. The mentalists instead try to count up how many people have the beliefs or knowledge to produce such artefacts, whether or not they are ever expressed. Behaviorism also frees memetics from defining a meme/host relationship, since artefacts in particular do not appear to have hosts, but propagate independently of their creators. The study of diffusion in behavioral practices or artefacts—long underway in the social sciences—can, according to behaviorists, serve as the proper empirical arm of memetics, which merely coats this standard endeavor in more explicitly evolutionary garb.

Behaviorists suggest that activities like making pots are the memetic equivalents of genotypes, while the mentalists would call such behaviors the phenotypic manifestations of memes-in-brains. This reversal of roles—thinking of behavior as the 'genotype' rather than 'phenotype' of culture—has some intuitive appeal. It is easy to think of spoken phrases as replicators—repeated, say, in a chain of people playing the game of Whispers. Similarly, the photocopying process can be seen as the replication of information embodied in ink-on-paper. However, this flipping of memotypes and phemotypes makes the behaviorist and mentalist positions potentially antithetical with respect to the essential theoretical distinction between replication and interaction. So even this brief foray into attempts at defining memes suggests there is disarray at a fundamental level in the subject.

What is culture?

The explanatory target of memetics, at least as narrowly conceived, is culture. Unfortunately, there is perhaps an equal amount of controversy about what culture might be as we have seen surrounding the concept of memes. Culture has been variously defined as a social construction, a 'text', social behaviours, artefacts or the mental entities (ideas/beliefs/values) in people's heads. Indeed, in the history of anthropology, there has been a good deal of controversy about what categories of things can be included in the definition of this central concept. As noted above, meme researchers tend to be cognitivists, restricting the notion to mental entities. But some memeticists would only include certain kinds of mentemes—arguing that emotions, for example, do not replicate, or are not infectious (e.g., Blackmore 1999).

A possibility which generally goes unrecognized by memeticists is that culture might be explained without recourse to memes at all. Some would argue that culture is just a new phenotypic strategy used by the most prominent class of replicators, genes (e.g., Flinn and Alexander 1982) rather than the product of a novel, quasi-independent class of replicators (memes) with their own interests (e.g., Brodie 1996; Lynch 1996). One of these theories is wrong: either memes exist or they do not.

Nevertheless, many researchers blithely discuss features of memes, ignoring the fact that their existence has yet to be proven. Most current discussion in memetics attempts to pin down the features of memes when there is as yet not even a standard codification of the concept (Rose 1998; Wilkins 1998). For example, Blackmore (1999) argues we can get some way without bothering about defining memes. The behaviorists, as I noted above, suggest that to make some progress we should ignore difficulties associated with the indefinable mental states associated with memes, and measure observables like behavior. Similarly, work in gene–culture coevolution (Boyd and Richerson 1985; Cavalli-Sforza and Feldman 1981; Durham 1991) is founded on the assumption of a quasi-independent line of cultural inheritance, and hence implies the existence of a cultural replicator. Models from this latter school indicate that natural selection can favor the transmission of acquired information and the persistence of social learning processes (e.g., Boyd and Richerson 1996). However, they do not prove that such abilities underlie human culture, nor that information packets with the characteristics of cultural replicators exist.

Surely, if memes exist, they must leave traces in the world. It seems that a firmer notion of what a meme is must precede any empirical search for them. While it is possible they will be found by accident, fortunes will surely be much brighter if foragers for memes have a clear 'search image' in place. In the absence of a well-founded model, recourse has simply been to argue from analogy to the best-known replicator, the gene, with little attention being paid to the necessity of identifying mechanisms for either replication, selection, variation or transmission. Many of the claims made about memes could be false because the analogy to genes has not proven productive. Memetics at present remains linked conceptually but not ontologically to biology.

Linking memes to culture

The vagueness of the meme concept naturally makes it difficult to find an appropriate way to link memes to culture. There are two main approaches to this problem. The first takes memes to be analogous to pathogens. Indeed, the literature of memetics is hugely infected with epidemiological terms—most readily seen in the titles of meme articles and books: 'virus of the mind' (Dawkins 1993; Brodie 1996), or 'thought contagion' (Lynch 1996). It is from epidemiology—traditionally a subject that takes a diffusionist perspective—that memetics gets its almost obsessional concern with the transmission of information. The main epidemiological question is: What factors influence the distribution or relative rate of spread of 'mind viruses' in a population? Qualities of memes themselves are typically viewed as determining their relative success in the replication stakes. But this makes it seem as if memeticists are simply saying that those memes are 'fittest' which survive and reproduce—which leads to a charge of tautology (Wilson 1999).

The second major strain of thought in memetics sees the meme primarily as a replicator. 'Replicator' is a notion coming from the same book in which the word 'meme' was itself coined: Dawkins' *The Selfish Gene*. A replicator is 'anything in the universe which interacts with its world, including other replicators, in such a way that copies of itself are made' (Dawkins 1978). In this neologism, Dawkins meant to emphasize that the evolutionary process identified by Darwin could be generalized to other substrates besides DNA—such as cultural information inherited through social transmission. In a similar fashion, Dawkins generalized the phenotype notion through use of the term 'vehicle', described most famously with reference to organisms as the vehicles which genes use to lumber around the environment. David Hull, a prominent philosopher of biology, soon thereafter modified the vehicle notion somewhat, to eliminate its implicit limitation to the case of phenotypic development. He adopted instead the term 'interactor'. Interactors are 'those entities that bias replication because of their relative success in coping with their environments' (Hull 1982: 316). This definition emphasizes the interactor's role as an ecological behavior-generator to achieve the differential copying of the replicator-based information it carries around. The replicator/interactor distinction is now standard in philosophical discussions of the evolutionary process, and reappears in many of the chapters that follow.

The theoretical foundation for the replicator analogy is evolutionary biology rather than epidemiology. The questions that come to the fore from this perspective are somewhat different as well: What are the mechanisms of heredity, selection, and mutation for memes? What is their origin? Although this arguably gives memetics a stronger theoretical foundation, the problem is that these questions are hard to answer.

Thus, we currently have at least two rival paradigms contending for dominance in memetics—the 'meme-as-germ' and 'meme-as-gene' schools. Their formal theories—epidemiology and population genetics—are equivalent at an elementary level (Cavalli-Sforza and Feldman 1981: 33). So strictly speaking, the diffusionist representation is based on the same three elements as evolutionism: innovation, selection, and reproduction. Nevertheless, the two schools have distinct intellectual histories, disciplinary agendas, and popular perceptions. This is largely due to the fact that epidemiology has not traditionally been concerned with the issues that are important from a theoretical evolutionary point of view, being a rather more pragmatic science with the clinical goal of curing disease. Where diffusionism primarily focuses on the spatial dimension of reproduction—or the geographical spread of a phenomenon—evolutionism focuses on the *temporal* dimension of reproduction—that is, on the continued existence and maintenance

of a phenomenon. Further, like its biological cousin, memetic epidemiology largely ignores how a 'virus' duplicates itself or mutates, regarding
innovation as a rare and unique occurrence. Identifying what the
selective forces on a pathogen might be is also not a high priority for
biological or cultural diffusionists, although they often work with
concepts such as barriers to diffusion and differences in susceptibility
(in memetic terms, receptivity to new ideas). And whereas evolutionists acknowledge the possibility that the same innovation can occur
several times at different places independently, the source of a variant
strain is typically not a concern to the epidemiologically minded.

However, such internecine arguments about the nature of memes
and culture belie a more general debate in the social sciences: whether
culture can be treated strictly as socially transmitted information in the
first place. While the idea that culture is somehow cognitive, or inside
the head, is now generally accepted, it it not universal. And even among
those who accept cognitivism in principle, some argue there are aspects
of culture which lie outside any individual head—for example, that
emergent social-structural qualities or material artefacts should be
included in the definition. Thus, the question arises: is culture amenable
to scientific investigation, and if so, is selectionism the most productive or congenial viewpoint to adopt? While assiduously eschewing the
'Social Darwinist' heritage, contemporary strains of evolutionary social
theorizing nevertheless speak of 'optimality' and 'adaptation', which
some see as disturbingly close to a panegyric for the social status quo.
As Dennett has suggested, perhaps a cultural replicator dynamics
produces more heat than light.

So several aspects of the standard memetic view, as it has thus far
developed, may be criticized. First, memes-as-replicators may not
discriminate the most important features of cultural traits. Culture may
not in fact be composed only of socially transmitted units of information—in effect, there may be no identifiable or measurable unit of
culture. Rather, culture might be considered—or at least felt to be—a
large, interconnected body of implicit knowledge which only has
meaning as a whole.

Second, cultural phenomena may be changed by forces other than
interactions among a set of mental replicators. This could be because
important components of culture are not in people's heads. Some argue
that at least some cultural phenomena are environmental (e.g., in the
form of artefacts), or emergent—a quality of human groups which is

constrained, but not strictly determined by, variation in beliefs and values among individuals.

Thus, disputes rage at three levels:

(1) whether culture is properly seen as composed of independently transmitted information units;

(2) whether these so-called memes have the necessary qualifications to serve as replicators; and

(3) whether a Darwinian or selectionist approach such as memetics is the most feasible or desirable form for a science of culture to take.

The objective of this book is to bring together the main contenders on this nested series of questions, both pro and con. Subsequent chapters thus present representative voices from the range of opinions currently available on the topic of memes.

Ways of seeing memes

The popularity of Susan Blackmore's recent book, *The Meme Machine*—together with Dennett's earlier advocacy (most notably in his book *Darwin's Dangerous Idea*)—has resulted in a substantial revival of interest in memes. Thus, it is appropriate that Blackmore presents in the first chapter a rousing defense of what might be called 'radical memetics'. This is the belief that memetic processes can explain a wide range of phenomena, including the rise of big brains, culture, consciousness, and notions of self. Blackmore here recounts and defends herself against some of the major criticisms of her book. These points of contention include seeing the evolution of the large human brain strictly as a response to the pressure of producing better memes, and the restriction of memetics to traits learned through imitation.

Perhaps the most important claim in Blackmore's work is the concept she calls 'memetic drive', which she believes is unique to the memetic perspective and distinguishes it from alternative evolutionary theories of culture, such as evolutionary psychology (e.g., Barkow *et al.* 1992) and gene–culture coevolutionary theory (e.g., Boyd and Richerson 1985). This drive is how the causal power of memes, derived from their ability to influence replication, manifests itself—primarily over the course of human evolution. This drive underlies most of the other claims Blackmore makes in her book (echoed here), particularly about the role of memes in explaining sociobiological conundrums. These evolutionary paradoxes include the hypertrophy of the human brain, the extravagance of human language (since much simpler communication systems are sufficient to organize other animal societies), and the tendency for humans to engage in altruistic acts, even in large groups of non-kin. She also deals with the provocative issue of whether memes are likely in the course of their further evolution to become replicators that no longer depend on human hosts. This inspiring—or perhaps frightening—vision of memetics is targeted from numerous directions by the authors of later chapters.

Next, David Hull presents his personal view of what contemporary philosophy of biology has to say about memes-as-replicators. In the process, Hull makes a number of fundamental observations. For example, he demolishes the familiar misconception that cultural evolution is always faster than genetic change. What about the case of HIV, which mutates into a quasi-species within a single host's body within a period of months? In contrast, the theory of evolution still has not succeeded in colonizing many hosts in any form.

Hull also believes that memetics cannot rightfully be charged with Lamarckism—or the inheritance of acquired characteristics—because memes are defined as replicators, not interactors. As Hull contends, memes are analogous to genes, not phenotypic characteristics. From the perspective of genes, things like mental states or words are phenotypes, but this is irrelevant. From the memetic perspective, hearing words is acquiring memes, and hence becoming host to a new replicator. Passing along memes is therefore a Darwinian, not Lamarckian process. This highlights the importance of adopting the proper perspective—the 'memes' eye view'—when positing novel evolutionary processes.

Although generally sympathetic to memes, Hull takes issue—as do others who follow (see the chapters by Plotkin, Conte and Laland and Odling-Smee)—with Blackmore's restriction of memetics to 'information learned through imitation'. In her view, this is the only mechanism leading to descent with modification, and hence the only mechanism for social transmission which can properly be seen as evolutionary. Hull argues that this restricts memetics, unlike other evolutionary theories, to a single species—humans. While this leaves memetics of interest to us, it means that memes cannot play a role in explaining more general evolutionary trends like the increase in intelligence within some animal families.

However, Hull's main objective seems to be to use his magisterial voice to argue we should 'just get on with it'. As someone who has studied empirically the question of how science progresses, memeticists would probably do well take his advice to heart. Leave definitional issues until later, Hull declares, and concentrate on getting results. These should, in dialectical fashion, make theoretical questions more clear. In the same vein, Hull is careful to promote memetics directly: he cites the younger memeticsts who remain largely unacknowledged by the academic mainstream—due in some cases to lack of institutional affiliation and credentials. As he knows from his own studies of citation practices, this is a powerful way to help the eventual success of an upstart research program.

Our next contributor is the psychologist Henry Plotkin. He is particularly keen to assuage the fear implicitly underlying most social scientists' rejection of memetics (see the chapters by Kuper and Bloch): that it is yet another brand of biological domination. He cogently argues against memetics as a science that reduces culture to biology. This is because large-brained creatures like humans do not have enough genes to specify the connections established between their many neurons. As a result, the state of the brain largely reflects information-processing due to environmental pressures, including social stimuli, rather than genes. Further, since culture is the emergent result of big-brained creatures interacting with one another, there must be an additional level of complexity to the explanation of such a population-level phenomenon. This takes us far from genetic determinism in Plotkin's view.

Plotkin also identifies two kinds of memes, which he calls 'surface-level' and 'deep-level,' depending on the breadth or depth of knowledge structure they subsume. Deep memes, he argues, are not acquired through a single act of imitation, but rather through the integration of many experiences and perceptions. Plotkin hopes the notion of deep memes will assuage the fears of those who think that memetics is too atomistic to account for the learning of complex knowledge structures. (To the minds of these critics—represented here by Kuper and Bloch—not all knowledge acquired through enculturation is like the classic memetic examples of tunes and catch-phrases.) In good evolutionary psychological fashion, Plotkin suggests that deeply structured memes are likely to be the result of naturally selected modules in the brain. So presumably the commonality of deep memes is due at least in part to the universal psychological mechanisms of construction that Sperber (in a later chapter) talks about. This knowledge must therefore be distinguished from transmitted information *sensu strictu*. At the same time, some higher level functions of the brain (such as Plotkin's example, the supervisory attentional system) involve multiple domains. Presumably, deep memes result from the activity of such cross-level and cross-domain functions. Whether the distinction between surface and deep memes will hold up under empirical scrutiny, however, remains to be established.

The main concern of Rosaria Conte, in her chapter, is also to emphasize that memetics must be placed on a firm psychological foundation. Although she has this desire in common with Henry Plotkin, her preferred foundations differ from his. Rosaria Conte is among the modelers of cultural evolution. However, her tradition is not gene–culture coevolutionary theory (derived from the population genetics formalism), as in the case of the two pairings of, Laland and Odling Smee and Boyd and Richerson (discussed below). Instead, she is at the forefront of a movement in cognitive science to bridge the traditional concerns of agent-based modeling in computer science with human social psychology. In particular, she is less interested in analytic modeling than in simulation, especially computer-based simulations of complex agents in 'artificial societies'.

Conte's crucial claim is that memetics is necessarily restricted to intentional agents. The standard view, largely inspired by evolutionary biology, suggests that 'cognitively impaired' agents (such as lower animals) can transmit memes. But in Conte's view, memetics must be based on autonomous agents with decision-making abilities, summarized in her notion of a 'memetic agent'. In Conte's vision, memes can be almost any symbolic token, whether in minds or the environment (see her definition of meme near the close of her chapter). In this, she is quite far from standard evolutionary memetics, which would argue that there are many kinds of representations—even symbolic ones—which do not qualify as memes because they have no mechanisms for replicating themselves. But Conte can take a rather general view of memes because for her replication is the responsibility of the memetic agent. As the name suggests, such an agent is the primary mover in her system, not memes themselves. For Conte, memes do not have to be clever; rather, meme receivers or interpreters do. This is a point to which other contributors return.

Two controversial claims derive from Conte's central argument: that neither communication nor imitation is necessary for memetic transmission to occur. First, memes can be transmitted without true communication. For example, one can use deception, where the message is intended to modify the mental states of others (i.e., a meme is passed), but in such a way (if the deception is effective) that one's true intention is not communicated. Conte provides the example of leaving a light on in the house to deter burglars while one is away.

Second, memes can also diffuse through a population without overt imitation. For example, thanks to the preference to be like some elite class, individuals can seek to differentiate themselves by maintaining their elite traits—but only so long as they are rare. In effect, such memetic agents adopt traits *unlike* those modeled for them by others.

Thus, Conte would have us distinguish between kinds of transmission processes, depending on the psychological abilities of the sender and receiver. For her, a transmission process can be considered memetic when the sender and receiver of messages are able to manipulate each other's minds effectively, producing more stable traditions of information exchange. To determine whether a transmission process is memetic or not, then, we should always ask: do the sender and receiver have intentional states—that is, the ability to simulate the intentional states of others? In her view, social cognition matters because these abilities can lead to different social dynamics.

Some in memetic circles would argue that this unnecessarily restricts the kinds of agents which can be counted as memetic. In particular, it limits memes to the few species capable of intentional behavior. So the minimum requirements for memetic transmission are high in terms of the cognitive capacities of the sender and receiver, but low in terms of the symbolic content of the meme itself and with respect to the sophistication of transmission mechanisms. Conte is thus one who would psychologize memetics to a degree not seen elsewhere in this volume.

She also points out that while the memetic literature places considerable emphasis on beliefs, other kinds of mental states can be transmitted through social interaction as well—and perhaps with greater fidelity. The importance of how a meme is mentally represented lies in the fact that beliefs are not the same things as obligations, for instance, and this has implications for transmission parameters. In fact, Conte focuses almost exclusively on the case of norms. Norms are, for her, particularly interesting forms of memes because they have unique psychological qualities which influence their likelihood and direction of transmission compared to other forms of mentally represented information.

Kevin Laland and John Odling-Smee, in a rich chapter, argue that the developing vision of memetic transmission must be supplemented by an important process they call 'niche construction'. This is a process in which organisms, through perhaps instinctive behaviors such as building nests or merely excreting detritus, manipulate environmental factors which subsequently introduce important new selection pressures on them, as well as other species which interact with those new features of the environment. If these modifications persist, there can be feedback between the activities of one generation and the selective environments of the next. Laland and Odling-Smee call this transmission of modified environments 'ecological inheritance'. Models that include ecological inheritance, largely constructed by these same authors, have shown such feedback can produce novel evolutionary dynamics, and so should be considered when organisms construct their niches. Since the idea that this kind of activity has evolutionary importance adds an extra degree of complexity to evolutionary models, is unfamiliar, and remains controversial, Laland and Odling-Smee are at pains to present the case for including this complication in standard evolutionary theorizing.

They also present a novel theory of the evolution of the cultural capacity during the emergence of the hominid line. Laland and Odling-Smee's approach is founded on a conjunction of transmission vectors, ecological inheritance, and the accumulation of constructed features in the hominid niche. Their theory is at odds with Blackmore's take on the same topic (presented more fully in her book *The Meme Machine*), which involves sexual selection for imitative ability. As these contesting theories of the evolution of culture imply, different features of human psychology should be important for cultural transmission. An empirical contest between these competing theories should therefore be possible, at least in principle.

Like others before them, Laland and Odling-Smee provide a powerful argument in favor of opening up memetics to non-imitative social learning, and hence admitting non-human species to the memetic brotherhood. Laland and Odling-Smee thus differ with Blackmore's approach to memetics in several fundamental respects. This is a vivid demonstration of the multiple visions about even basic propositions among proponents within the memetic brotherhood.

The biologists Robert Boyd and Peter Richerson are rather more critical of the meme notion. They argue that memeticists have been far too enamored of one of Darwin's conceptual advances: the identification of natural selection as the mechanism for cumulative adaptation. They would convince us that Darwin's *other* great contribution—what Ernst Mayr calls 'population thinking'—is a more appropriate organizing principle for an evolutionary theory of culture. This is because, in their view, cultural evolution need not involve selection among replicators. Culture can instead be considered a pool of information that is passed to subsequent generations via a variety of hypothetical mechanisms which do not resemble their biological counterpart, natural selection on genes. For example, if one allows selection to take place at multiple levels of organization, continuity of cultural traditions can be produced without information being passed from individual to individual. Instead, the alternatives generated by individual learning which survive can be constrained by mechanisms operating at the group level. The result however, is what we observe: the regularity of cultural traits being preserved over time. Alternatively, individuals may average the values of what they learn from others, but then also internally generate variants on this average through their own cogitations. If these variation-reducing and variation-augmenting processes balance each other out, there can be a high degree of correlation in what different generations believe. This is again the heritability of cultural traits without the replication of specific bits of information. Since heritability is only concerned with correlations, not mechanisms, these scenarios fall within the domain of evolutionary processes, without being based on replication in the same fashion as genetic inheritance.

This is a strong stance to take, but one which is forcefully argued, impeccably logical, and aptly illustrated with empirical examples. Hull (this volume) counters that 'any adequate understanding of selection ... requires the specification of the mechanisms that are bringing about these correlations' in cultural features between generations. He suggests that no mechanism besides inheritance through descent is currently known to have the necessary qualities to sustain an evolutionary process. Nevertheless, Boyd and Richerson's hypothetical mechanisms are consistent with their formal modeling in the population genetics-based gene–culture coevolutionary theoretical tradition. The threat of this logical possibility to memetics is therefore real.

Boyd and Richerson also draw attention to the fact that both genetic and cultural transmission are likely to play a role in the continuity of traditions: unlike most memeticists, they model *dual* inheritance. So evolutionary psychology—the genetic transmission of predispositions for interpreting inputs, or for the ability to imitate itself—is accounted for in their approach to cultural evolution. It is more general than memetics, Boyd and Richerson claim, because it is not specific to the standard memetic assumption of *particulate* inheritance.

Along the way, Boyd and Richerson also provide a devastating critique of the evolutionary psychological notion that human culture can be almost exclusively innate. Their point is that cultural innovations, such as technology simply accumulate information faster than is possible through genetic inheritance. They rehearse the now standard argument that what separates human culture from protoculture in other species is the same ability to accumulate innovations across generations. The young of other species only manage to reinvent their parents' wheels before dying themselves, thus merely reproducing what earlier generations have bequeathed to them. They end with an encomium about the ability of their population-based approach to reconcile the social sciences with its individual-based cousins such as economics and psychology. May their wish come true!

Dan Sperber, in a compelling contribution, sets up a major empirical hurtle for any future discipline of memetics. Sperber's key point is that one can observe very similar copies of some cultural item, link these copies through a causal chain of events which faithfully reproduced those items, and nevertheless not have an example of memetic inheritance. This is because each copy of the item may have been produced by following 'local' instructions, rather than a blueprint received (typically in the form of a message) from the previous producer in the causal chain. The result may be similar beliefs, behaviors or artefacts, but the process is not one of copying. What matters is where the instructions come from: true inheritance requires that the information which makes the items similar be acquired from the original. As Sperber notes, many discussions in memetics do not distinguish between similarity which arises from reproduction and from inheritance. Causation and similarity are not enough. One must also have the relevant information being passed down the causal chain for true evolutionary replication.

Sperber's argument puts some flesh on the bones of Boyd and Richerson's contention that cultural evolution can logically proceed without replication. Sperber suggests this is not just idle speculation, but often the case. Based largely on his own work in human (linguistic) communication (see Sperber and Wilson 1995), Sperber asserts that the kind of high fidelity copying memeticists assume is characteristic of cultural transmission will only ever be a small proportion of cultural learning. It is but the limiting case of a much more complex process involving multiple steps of inferencing—first, to establish the sender's intention, and second, based on that, to decode what the message means. Since words and other linguistic units are memeticists' favorite example of memes (considered as culturally transmitted particles), Sperber's critique is significant. As he concludes (this volume), 'memeticists have to give empirical evidence to support the claim that, in the micro-processes of cultural transmission, elements of culture inherit all or nearly all their relevant properties from other elements of culture that they replicate'. His position is closely allied to the idea in evolutionary psychology that most of culture represents innate responses evoked by particular circumstances, rather than information transmitted pheno-typically between individuals (Tooby and Cosmides 1992).

While the other chapters concentrate on making the notion of meme clearer and sharper, Adam Kuper, in the penultimate contribution, suggests that the target memetics is seeking to explain—culture—is itself fuzzy. He even goes so far as to suggest that culture does not even exist in any meaningful sense. This makes the memetic project rather like a blunt arrow shooting into the dark. At the very least, it renders the memetic project less likely to succeed. Culture has come to be considered such a conflation of disparate entities, such an all-enveloping *Weltanschauung*—the very fabric of everyday life—that it becomes difficult to tease it apart in the ways memetic analysis would require.

Kuper also draws some lessons from history. He points out that culture used to be associated with the aristocratic notion of 'civilized taste', but now commonly connotes 'shared beliefs'. Culture began as the thing which distinguishes us from animals (a distinction that has becoming increasingly blurred, in particular as we have learned more about other primates). Now, it has become the Boasian notion of what distinguishes one human group from another, each culture being equally good and valuable. So culture-*qua*-civilization becomes culture as an accumulating inheritance of ideas, practises, and institutions. In effect, the concept of culture has been democratized, to reflect current political sensibilities. The memetic perspective, of course, depends on its explanatory target, culture, having this newer, diffusionist feel, because the memetic idea is that ideas spread epidemiologically like viruses. This analogy to viruses brings culture closer to biology. But this proximity of a neighboring discipline to the anthropological home turf is just what makes Kuper nervous, as readers will see. It draws up specters from earlier times in the history of the social sciences which do not sit well in memory.

Finally, Maurice Bloch, a social anthropologist like Adam Kuper, is favorably disposed toward the basic idea of transmitted culture (as he makes clear in his chapter title). Nevertheless, he complains bitterly about the ignorance displayed by memeticists of the rich academic literature on the topic of cultural change. This ignorance is galling to those who study culture professionally—to wit, sociocultural anthropologists like Bloch. As he is at pains to point out, this history is largely news to those approaching culture from other disciplines—and most memeticists are either from 'hard science' or psychology backgrounds. But their ignorance, particularly of cultural anthropology, is not excusable because they are explicitly attempting to explain the central concept in that discipline—culture.

This ignorance also leads memeticists to fall into traps already recognized and currently avoided by theoretic traditions in the social sciences without a biological pedigree. Like Kuper, Bloch takes an historical look at anthropological theory to make his argument. In particular, he likens memeticists to the diffusionists who briefly held sway at the beginning of the twentieth century, and reviews criticisms against diffusionism. Like Sperber and Kuper, he argues that considering cultural traits to be separable and independent bits of information flitting through populations in a carefree fashion is not an accurate description of ethnographic reality. As Bloch (this volume) puts it: 'The problem which anthropologists immediately recognise with memes lies . . . [with] the notion that culture is ultimately made of distinguishable units which have "a life of their own". Only then would it make sense to argue that the development of culture is to be explained in terms of the reproductive success of these units "from the memes point of view"'. Bloch also emphasizes the importance of Sperber's primary critique of the meme notion, suggesting that even if cultural traits take on particulate form during transmission, they nevertheless undergo substantial reformulation as they are integrated by individuals into their knowledge bases. Communication involves not just transmission, but the recreation, or reconstruction, of information by recipients.

The essential claim by Kuper and Bloch, then, is that culture is not divisible into units because it is a complex, heterogeneous thing. Others 'inside' the evolutionist fold agree with them in this respect—in particular, Boyd and Richerson, and perhaps Sperber. So, a central problem for memetics is obviously to begin to isolate and identify these 'bits' of culture. Perhaps only through such an identification will the utility of this approach be broadly accepted in social scientific circles.

Conclusion

This brief review should make it clear that a variety of stances can be legitimately taken with respect to the notion of memes—or at least the current implementation of the notion. In fact, there remains considerable disagreement about the value of memes, as will become evident to even the idle reader. From whence does this disgruntlement spring? From intrinsic defects in the notion (thus kneecapping any future development of the field from its incipient state), in incidental features of its present manifestation, or from intellectual agendas having little to do with memetics itself? The reader must judge.

At minimum, the following dialogue establishes areas of common ground, as well as highlighting the points of remaining contention. It is designed to represent the state of debate on the utility of memes as the foundation for the study of culture, and it is hoped, sets the terms for future discussion about the possibility of a Darwinian science of culture.

References

Barkow, J. H., Cosmides, L., and Tooby, J. (ed.) (1992). *The adapted mind*. Oxford: Oxford University Press.

Benzon, W. (1996). Culture as an evolutionary arena. *Journal of Social and Evolutionary Systems*, **19**: 321–362. [http://www.newsavanna.com/wlb/CE/Arena/Arena00.shtml]

Blackmore, S. (1999). *The meme machine*. Oxford: Oxford University Press.

Boyd, R. and Richerson, P. J. (1985). *Culture and the evolutionary process*. Chicago: University of Chicago Press.

Boyd, R. and Richerson, P. J. (1996). Why culture is common, but cultural evolution is rare. *Proceedings of the British Academy*, **88**: 77–93.

Brodie, R. (1996). *Virus of the mind: The new science of the meme*. Seattle: Integral Press.

Callebaut, W. and Pinxten, R. (ed.) (1987). *Evolutionary epistemology: A multiparadigm program with a complete evolutionary epistemology bibliography*. Dortrecht: Reidel.

Carroll, J. (1995). *Evolution and literary theory*. Columbia: University of Missouri Press.

Cavalli-Sforza, L. L. and Feldman, M. W. (1981). *Cultural transmission and evolution: A quantitative approach.* Princeton: Princeton University Press.

Cziko, G. (1995). *Without miracles: Universal selection theory and the second darwinian revolution.* Cambridge, MA: MIT Press.

Dawkins, R. (1976). *The selfish gene.* Oxford: Oxford University Press.

Dawkins, R. (1978). Replicator selection and the extended phenotype. *Zeitschrift für Tierpsychologie,* **47**: 61–76.

Dawkins, R. (1982). *The extended phenotype.* Oxford: Oxford University Press.

Dawkins, R. (1987). *The blind watchmaker.* New York: Norton.

Dawkins, R. (1993). Viruses of the mind. In *Dennett and his critics: Demystifying mind,* (ed. B. Dahlbom), pp. 13–27. Oxford: Blackwell.

Dennett, D. (1995). *Darwin's dangerous idea.* London: Penguin.

Durham, W. H. (1991). *Coevolution: Genes, culture and human diversity.* Stanford: Stanford University Press.

Flinn, M. V. and Alexander, R. D. (1982). Culture theory: The developing synthesis from biology. *Human Ecology,* **10**: 383–400.

Gardner, M. (2000). Kilroy was here [Review of *The meme machine* by Susan J. Blackmore]. *Los Angeles Times,* 5 March.

Gatherer, D. G. (1998). Why the thought contagion metaphor is retarding the progress of memetics. *Journal of Memetics–Evolutionary Models of Information Transmission,* **2**. [http: //www.cpm.mmu.ac.uk/jom-emit/1998/vol2/gatherer_d.html].

Gatherer, D. G. (1999). Reply to commentaries. *Journal of Memetics–Evolutionary Models of Information Transmission,* **3**. [http: //www.cpm.mmu.ac.uk/jom-emit/1999/vol3/gatherer_reply.html].

Hull, D. L. (1982). The naked meme. In *Development and culture: Essays in evolutionary epistemology* (ed. H. C. Plotkin), pp. 272–327. Chichester: Wiley.

Koza, J. R. (1992). *Genetic programming: On the programming of computers by means of natural selection.* Cambridge, MA: MIT Press.

Krebs, J. R. and Davies, N. R. (1997). *Behavioural ecology: An evolutionary approach.* Oxford: Blackwell.

Lakatos, I. (1970). The methodology of scientific research programmes. In *Criticism and the growth of knowledge* (ed. I. Lakatos and A. Musgrave), pp. 91–196. Cambridge: Cambridge University Press.

Lake, M. (1999). Digging for memes: The role of material objects in cultural evolution. In *Cognition and material culture: The archaeology of symbolic storage* (ed. C. Renfrew and C. Scarre). Cambridge: McDonald Institute for Archaeological Research.

Lanier, J. (1999). On Daniel C. Dennett's 'The evolution of culture'. *Edge,* **53**, 8 April. [http: //www.edge.org/documents/archive/edge53.html].

Lynch, A. (1996). *Thought contagion: How belief spreads through society: The new science of memes.* New York: Basic Books.

Lynch, A. (1998). Units, events and dynamics in memetic evolution. *Journal of Memetics–Evolutionary Models of Information Transmission,* **2**. [http: //www.cpm.mmu.ac.uk/jom-emit/1998/vol2/lynch_a.html].

McGuire, M. T. and Troisi, A. (1998). *Darwinian psychiatry.* New York: Oxford University Press.

Nelson, R. R. and Winter, S. G. (1982). *An evolutionary theory of economic change.* Cambridge, MA: Harvard University Press.

Nesse, R. M. and Williams, G. C. (1994). *Why we get sick.* New York: Random House.

Pinker, S. (1994). *The language instinct: the new science of language and mind.* London: Penguin.

Rose, N. (1998). Controversies in meme theory. *Journal of Memetics–Evolutionary Models of Information Transmission,* **2.** [http: //www.cpm.mmu.ac.uk/jom-emit/1998/ vol2/rose_n.html]

Smolin, L. (1997). *The life of the cosmos.* London: Weidenfeld & Nicolson.

Sperber, D. and Wilson, D. (1995). *Relevance: Communication and cognition* (2nd edn). Oxford: Blackwell.

Tooby, J. and Cosmides, L. (1992). The psychological foundations of culture. In *The adapted mind* (ed. J. H. Barkow, L. Cosmides, and J. Tooby), pp. 19–136. Oxford: Oxford University Press.

Westoby, A. (1996). *The ecology of intentions: How to make memes and influence people: Culturology.* Boston: Center for Cognitive Studies.

Wilkins, J. S. (1998) What's in a meme? Reflections from the perspective of the history and philosophy of evolutionary biology. *Journal of Memetics–Evolutionary Models of Information Transmission,* **2.** [http://www.cpm.mmu.ac.uk/jom-emit/1998/vol2/ wilkins_js.html]

Wilson, S. R. and Czarnik, A. W., (eds) (1997). *Combinatorial Chemistry: Synthesis and Application.* New York: John Wiley and Sons.

Wilson, D. S. (1999). Flying over uncharted territory [Review of *The meme machine* by Susan Blackmore]. *Science,* **285:** 206.

CHAPTER 2

The memes' eye view

Susan Blackmore

The new replicators

Robert Aunger, in his Introduction, has suggested that memeticists such as myself face a challenge: either to provide an existence proof for memes, or to come up with supported, unique predictions from meme theory. I suggest, however, that no existence proof is required and we would do better to concentrate on whether meme theory can be of any scientific value or not.

The reason no existence proof is required is the way 'meme' is defined. When Dawkins (1976) first coined the term he wanted an example of a replicator other than the gene. He based the name for his new cultural replicator on the Greek word *mimeme*—meaning that which is imitated. He intended imitation 'in the broad sense' (a point to which I shall return) but was very clear that whatever is passed on when people imitate each other—that is the meme. This clarity is reflected in the new Oxford English Dictionary definition of 'meme (mi: m), n. Biol. (shortened from mimeme . . . that which is imitated, after GENE n.) An element of a culture that may be considered to be passed on by non-genetic means, esp. imitation'. Although many authors use widely differing definitions I suggest we stick to this simple one. Doing so avoids many problems. It also becomes clear why no existence proof is required. As long as we accept that people do, in fact, imitate each other, and that information of some kind is passed on when they do then, by definition, memes exist.

We might, however, be a little more strict in our requirements and demand that memes must be shown to be replicators to count as existing. To be a replicator something must be capable of sustaining the evolutionary process of heredity, variation, and selection (Dawkins

1976) or blind variation with selective retention (Campbell 1960). It must, as Dennett puts it, undergo the evolutionary algorithm—that blind, mechanical procedure which creates 'Design out of Chaos without the aid of Mind' (Dennett 1995: 50).

Whichever scheme you prefer, memes fit. By definition they are inherited because they are passed on by imitation. They undergo selection in the sense that people are exposed to far more memes than they can possibly remember, let alone pass on again. And memes vary, whether by degradation (as occurs with errors of perception, memory, or reconstruction) or by creative recombination (as when different memes are put together to produce new combinations). The former is not helpful to memetic evolution in that memes are likely to lose any of the 'good tricks' they have accumulated (Dennett 1995). Recombination should be a more effective way of producing viable memes that will outperform those produced by degradative variation. In any case, memes clearly vary and therefore fit neatly into the evolutionary algorithm. In other words, memes are replicators. The importance of this is that replicators are the ultimate beneficiaries of any evolutionary process. Dennett (1995) urges us always to ask *cui bono*? or who benefits? and the answer is the replicators. This means that if we have a new replicator—the meme—there is a new entity whose interests must be taken into account.

I suggest that no further proof of the existence of memes is required.

The interesting question then, is not whether memes exist, but whether taking the memes' point of view can lead to any useful scientific work. In other words, is memetics a worthwhile endeavour? I believe that it is—not just because I am enjoying the startling new meme's eye vision of the world—but because memetics provides new solutions to old problems, among them the origins of our large brain with its specialised language and unique intelligence.

Why the emphasis on imitation?

Before I discuss the advantages of the memetic perspective, I would like to consider one further issue associated with the definition of memes. I have chosen to stick to Dawkins's original formulation of memes as information that is passed on by imitation. Others differ here. For example, Cavalli-Sforza and Feldman (1981) base their model of cultural transmission on traits that can be passed on by imprinting, conditioning, observation, imitation or direct teaching. Durham's (1991) coevolutionary model refers to both imitation and learning. Runciman (1998) refers to memes as instructions affecting phenotype passed on by both imitation and learning. Laland and Odling-Smee (this volume) argue that all forms of social learning are potentially capable of propagating memes. Among meme theorists Brodie (1996) includes all conditioning as memetic, and Gabora (1997) counts all mental representations as memes regardless of how they are acquired.

My reason for restricting meme acquisition to imitation (i.e., excluding other kinds of learning) is my suspicion that only imitation is capable of sustaining a true evolutionary process (Blackmore, in press). In individual learning (such as imprinting, classical conditioning, and operant conditioning), nothing is copied from one individual to another, so there is no basis for a replicator to operate. In other forms of social learning, such as stimulus enhancement or local enhancement, the behaviour of two individuals is involved and that of the learner ends up similar to that of the original performer, but the behaviour is not copied from one individual to another. For example, cultural traditions such as tits learning to open milk bottles or chimpanzees using sticks to fish for termites, are thought to spread by stimulus enhancement. Each individual learns the skill anew, having had its attention drawn to the location, the available materials or the stimulus of a pecked bottle top. In such traditions, as Tomasello *et al.* (1993) point out, there is no accumulation of modifications over generations—no cultural ratchet effect. Similarly, Boyd and Richerson (this volume) argue that only observational learning of novel behaviours allows cumulative cultural change.

Jablonka (1999) provides the useful distinction between reproduction and replication of behaviours. You could say that in other forms of social learning the same behaviour is apparently *reproduced* (such as washing sweet potatoes or pecking at milk bottle tops), but it is not *replicated*—that is, copied. This means there is no opportunity for variations on the copied behaviours to compete with each other, for truly novel behaviours to spread, or for cumulative change. In other words, without imitation there is no replicator and no new evolutionary process.

To some extent, this difference could be seen as an issue of copying fidelity. You could argue that other forms of social learning can reproduce new behaviours with sufficiently high fidelity to count as replication and to sustain evolution. This is an empirical question worth researching if these issues are to be resolved (Blackmore, in press). The question would be which kinds of social learning can reproduce behaviours with sufficient fidelity to maintain them intact over several generations of copying, and to allow for selection between variants and for cumulative change. Such research may reveal that in fact other kinds of social learning can sustain such an evolutionary process, in which case they should be included as processes that replicate memes. However, working without such information and with the current uncertainties over definitions, I would argue that only imitation has the capacity to sustain an evolutionary process and this is a good reason for restricting the definition of memes to that which is imitated.

There is also the related question of whether you choose to apply the word 'culture' to behaviours that are spread by other forms of social learning. If you do, then some monkeys, rats and birds have culture but, by my definition at least, they do not have memes. On the other hand, dolphins, some songbirds, and possibly elephants and chimpanzees do have memes because they are capable (at least to some extent) of copying novel behaviours or sounds by imitation.

A different question arises at the other end of the scale when we think about memes passed on by complex human processes such as reading, writing, and direct instruction. I presume that Dawkins meant to include these when he used the phrase 'imitation in the broad sense'. We may not wish to count these as forms of imitation, but I would argue that they build on the ability to imitate and could not occur without it. Learning language requires the ability to imitate sounds, and instructed learning and collaborative learning emerge later in human development than does imitation (Tomasello *et al.*, 1993). All these complex human skills clearly entail the copying of information from one person to another. Variation is introduced both by degradation due to failures of human memory and communication, and by the creative recombination of different memes. Selection occurs because of the limitations on available communication channels, time, memory, and other sorts of storage space. Information passed on by these means therefore fits the evolutionary algorithm.

A final problem concerns creativity. Many people seem to think that imitation is a crude and blindly mechanistic process that is the antithesis of human creativity, which is conscious and purposive. Theirs is indeed a very different view from my own, and entirely misses the point that evolutionary processes are creative—arguably the only creative processes on the planet. The alternative, first sketched out by Campbell (1960), is that just as biological creations come about through natural selection, so human artistic, literary, and scientific creations come about through memetic selection. In both cases the creative force is the evolutionary algorithm. Human achievements are no less creative for that, but our own role has to be seen as that of the clever imitation machine taking part in this new evolutionary process, rather than a conscious entity who can stand outside of it and direct it.

The human brain

I suggest that memetics can provide a new explanation for the origins and evolution of the human brain. Since memes are, by definition, passed on by imitation, they must have first appeared when our ancestors became capable of imitation. This would have made an enormous difference in evolutionary terms because memes were a new replicator that started evolving in their own way and for their own replicative ends. Since that time human evolution has been driven by two replicators, not just one. I suggest that this is why humans are unique. It was the advent of the new replicator that changed the ground rules forever. Since then, meme–gene coevolution has produced the enormous human brain which is designed not just for the benefit of genes, but for the propagation of memes.

The sheer size of the human brain requires some kind of evolutionary explanation. It is roughly three times larger than would be expected for an ape of our size and weight. It uses a prodigious amount of energy both to produce and to run, and is dangerous to give birth to. Not only is it unusually large but it has been restructured in various ways and apparently specially adapted for the production and comprehension of language.

Early theories to explain the big brain focused on hunting and foraging skills, but their predictions have not generally held up, so more recent theories have emphasized the complex demands of the social environment (Barton and Dunbar 1997). Chimpanzees live in complex social groups and it seems likely that our common ancestors did too. Making and breaking alliances, remembering who is who to maintain reciprocal altruism, and outwitting others, all require complex, fast decision-making and good memory. The 'Machiavellian Hypothesis' emphasizes the importance of deception and scheming in social life and suggests that much of human intelligence has social origins (Byrne and Whiten 1988; Whiten and Byrne 1997). Dunbar (1996) argues that gossip is the human equivalent of grooming in that it enables large social groups with complex relationships and reciprocal altruism to be maintained. This, he argues, explains the evolutionary advantage of language, and the need for language drove the increase in brain size.

Most of these theories entail gradual changes in abilities and brain size, but others include one or more transitions. For example, Donald (1991) proposes a three stage account of how human brains, culture, and cognition coevolved. His first step is a 'revolution in motor skill' (Donald 1993: 739) that he calls 'mimetic skill'. He argues that the anatomical changes necessary to support speech evolved in a mutually reinforcing manner with the lexical capacity. However, his term 'mimetic' is quite unrelated to the term 'memetic'. By mimesis he means 'the ability to produce conscious, self-initiated, representational acts that are intentional but not linguistic' (Donald 1991: 168). He specifically excludes 'simple imitative acts' and concentrates on the importance of representation—whether externally to someone else or internally to oneself. This emphasis on symbolic representation makes Donald's theory quite different from the one proposed here, which rests entirely on the premise that copying actions from one individual to another creates a new replicator. Whether those actions represent anything or are symbolic is quite irrelevant to their role as replicators. Also Donald's theory, like most other theories of human evolution, ignores the possibility of a second replicator, and treats all adaptations as ultimately for the benefit of genes.

One possible exception is Deacon's (1997) theory of the coevolution of language and the human brain. Deacon argues that once simple languages appeared, they created selection pressure for bigger and better brains able to understand them. Although he does not use the term 'meme', he likens language to a parasitic organism with some of its features evolved for the purpose of passing the language on from host to host, even at the expense of the host's adaptations. He refers to the symbolic adaptation as a 'mind virus' that has turned us into the means for its own propagation (Deacon 1997: 436). However, his theory differs from the one I am proposing here in that the critical turning point was not the appearance of imitation but the point at which our ancestors crossed the 'symbolic threshold'. For Deacon symbolic refer-ence provided the only conceivable selection pressure for the evolution of hominid brains.

These various theories differ in many other ways but most share the conventional neo-Darwinian assumption: that the human brain was designed by evolution for the benefit of genes. In other words, their answer to Dennett's *cui bono?* question is genes. I propose, instead, that the human brain was primarily designed for the benefit of memes.

Memetic drive

I propose that there was a critical turning point in human evolution— when our ancestors acquired the ability to imitate. From this point on, memes started driving genes to produce a brain that was especially good at replicating those memes.

Imitation can be a 'good trick' from the genes' point of view because it reduces the costs of learning. We might liken imitation to stealing learned behaviour from someone else without having to take the risks involved, or put in the time and effort needed, to acquire it by trial and error or other forms of individual learning. Mathematical modelling has shown that social learning, including imitation, is worthwhile if the environment changes, but not too rapidly (Richerson and Boyd 1992). The point is that, although imitation might initially benefit the genes of the person who could do it, those genes had no foresight. They could not predict that they were letting loose a new replicator; one that need not 'be subservient to the old' (Dawkins 1976: 194).

Although speculating on the lives of our early ancestors is always dangerous, I would guess that the first memes were useful ones (i.e., useful to the genes), such as new ways of hunting or preparing food, ways of making baskets or simple tools, or dealing with social rela-tionships. However, once imitation was possible, memes could spread for many reasons other than their value to the genes that gave rise to them in the first place. So other not-so-useful memes would soon begin to exploit the new copying machinery and spread by imitation as well. These might include rituals, body decoration, burial rites, or music. In even a simple culture like this we have the basic ingredients for what I have called 'memetic driving'.

The mechanism works like this. The people who are best at imitation have an advantage over the rest because they can most easily acquire any useful new skills or artefacts, and most easily put together old memes to produce new ones—we may call these people 'meme fountains'. As long as there is some genetic basis to what made them meme fountains in the first place, then genes for being good at imitation will tend to spread (on ordinary Darwinian principles). Assuming that imitation is a difficult skill that requires a bigger brain, we have a simple argument for the increase in human brain size—although thus far my argument is the same as many previous theories.

The next step is that once memes are around, everyone has to start making decisions about whom and what to imitate. In general, it will pay others to copy the meme fountains because they are more likely to have useful survival-related memes. This gives a further survival advantage to the meme fountains, and their genes, in terms of improved power and status. If there are genes for imitating the best imitators, these genes will also spread in the gene pool. However, this may mean copying a fancy headdress, or a pleasing song or dance, as well as a new way of making stone tools or baskets. Memetic evolution now gets under way with various kinds of dances, headdresses, and songs competing with each other to be copied.

We now have two effects operating. First, everyone gets gradually better at imitating the successful memes, which means that more and more memes are created and culture expands. Second, genes for the ability to copy meme fountains and their popular memes have an advantage and more people come to behave this way. But this now creates selection pressure for the ability to discriminate between useful and useless memes (i.e., useful or useless from the genes' point of view), because copying a popular meme just might prove fatal. As memes evolve in one direction or another, according to the outcome of memetic selection and the kinds of memes the meme fountains happen to be best at propagating, survival increasingly depends on the ability to choose which memes to copy and which to avoid.

This is essentially the basis of memetic drive. Memes compete with each other to be copied and the winners change the environment in which genes are selected. In this way, memes force genes to create a brain that is capable of selecting from the currently successful memes.

One final step to the argument is that for similar reasons there may be an advantage to mating with the meme fountains. Sexual selection may therefore add to the pressures on genes to produce brains capable of imitating the currently successful memes.

This opens the way for an explanation of how the brain has been designed for language and other specialized abilities. The argument depends on the power of the successful memes, so which are they? The answer, according to general principles of evolution, should be memes of high fidelity, fecundity, and longevity (Dawkins 1976). Language is a good way of creating memes with high fecundity and fidelity. For example, sound carries better than visual stimuli to several people at once. Sounds digitized into words can be copied with higher fidelity than continuously varying sounds. Using different word orders in different circumstances opens up more niches for memes to occupy and so on. For this general reason we should expect language memes to succeed in memetic evolution, and then memetic driving to cause the spread of the genes that make that language possible. That is, in an environment in which simple language is spreading memetically, the meme fountains will have the best command of the new language because they are good at imitation, while the people who cannot pick it up will be at a disadvantage in a way they never would have been before language appeared. In addition, those who are best at picking up the new language may be preferentially chosen as mates. For these reasons, any genes involved in the ability to copy the language will tend to spread. As the evolving language changes through memetic competition, so genes are forced to follow. On this argument, the function of language is to spread memes. The genes had no choice but to follow where the memes led and produce a brain that was not only as big as the genes could carry, but was designed especially for propagating memes through language.

Is this theory testable? Some of the assumptions on which it depends could be tested. For example, it assumes that imitation is a difficult skill that requires a lot of processing capacity. Brain scan studies might refute this if it turned out that imitation does not use large areas of the brain, or that it does not involve the evolutionarily newer parts of the human brain. Computer simulations and mathematical models are already being used to test whether memetic drive really could produce an increase in brain size. For example, Higgs (in press) has developed a model in which memes can have both positive and negative effects on the fitness of a population of individuals. He not only found that genes for imitative ability are selectively favoured but that imitative ability increases slowly until a rapid transition occurs, after which memes spread like an epidemic and there is a dramatic increase in imitative ability as well as mean fitness. Kendal and Laland (in press) have built on models from gene–culture coevolution theory and shown that the strategy of imitating enhanced imitators will spread under a wide variety of conditions, opening up new ways of testing these hypotheses.

Not just big but special

Since I first proposed this argument (Blackmore 1999) many colleagues and critics have raised difficulties or questions about the proposed process of meme–gene coevolution. In particular, some critics of my book *The Meme Machine* have gained the impression that I believe the human brain is an all-purpose meme machine, designed to copy any old memes, and that its size is the only mystery to be explained. This is clearly not so, for our copying is highly selective. From soon after birth infants imitate facial expressions, hand movements and so on, but they do not imitate just anything they see; their imitation is selective (Brugger and Bushnell 1999). Adults imitate speech, and certain kinds of actions and behaviours, but not others. In view of this criticism, I would like to make clear the implications of the memetic driving hypothesis.

The theory is meant to be an argument of the following general form. Once memes arise they evolve by memetic competition (high quality memes spread at the expense of low quality memes) and they evolve faster than the genes that made them possible in the first place. Meme fountains (who have all the useful memes as well as all the ones that have spread for other reasons) survive better both because they have more useful memes and because other people copy them, giving them added power and status. So the memes that succeed in memetic competition change the environment in which genes are selected, giving an advantage to genes which help a person imitate the currently successful memes—whatever those memes happen to be. In addition, meme fountains may be selectively chosen as mates, although this is not essential to the argument.

I used this argument to provide an explanation for the 'language instinct' or 'language organ' but this may have been a poor choice not least because it is such a contentious issue. I will therefore try the argument out on something less contentious—the human enjoyment of music. Why do we humans, apparently alone among animals, invest large amounts of time and energy in producing and enjoying complex music? It is difficult, if not impossible, to provide an answer based on advantage to human genes (Pinker 1997). Dennett (1999) imagines an early hominid, for no particular reason, just happening to bang on a log and enjoying the sound, and someone else coming along and copying it. For reasons to do with perceptual systems, memory, or features of the environment, some versions of the drumming, and then humming, are more infectious and spread at the expense of others. And so the process goes on, with the advantage being to the bangings, whistlings and hummings (i.e., the memes)—not necessarily to the hominids' genes. Dennett then speculates that females might be more receptive to the winning hums.

Note that the idea that culture evolved by sexual selection is not new. Miller (2000) has argued that human culture in general, and music and art in particular, are mainly a set of adaptations for courtship. He cites evidence that musicians and artists are predominantly male and at their most productive during young adulthood. His theory does not, however, involve memes and is therefore slightly different from that proposed here. The difference is this. On Miller's theory, the songs (or other productions) are the cultural displays that guide females in their choice of mates—comparable to the peacock's tail. Presumably the songs evolve only because of differential female choice. However, on Dennett's theory and on the memetic driving hypothesis proposed here, the songs themselves compete to be copied. This memetic competition takes place in both males and females; the outcome being determined by features of the songs themselves (e.g., how easy they are to sing or remember) and of the perceptual systems and vocal tracts of the people trying to copy them. Meme–meme competition thus determines the direction taken by the evolution of music, dance, art and literature, as well as sexual selection.

Exactly the same argument can be applied to religions. This is also a contentious topic and relates to Dawkins's (1993) suggestion that religions are viruses of the mind. He pointed out that some of the world's great religions may have spread not because they are true, or because they help anyone, but because they are successful memes— successful because they are essentially 'copy-me' instructions backed up with threats, promises, and ways of preventing their claims from being tested. Rather than discussing a vast memeplex like Roman Catholicism, we may take the simpler example of a ritual dance supposed to bring rain. The rain dance may, by chance, coincide with the advent of rain, and so be copied. If some meme fountain does a particularly flamboyant version this meme may be copied even more, outperforming other versions. Being powerful in this society (and hence acquiring a survival advantage) now becomes linked with being able to copy these winning dances. People not only copy these successful memes, but mate with the people who display them, so that any genes implicated in being good at these dances (or prayers, or fervent displays of belief in God, or singing hymns, or . . .) will tend to increase. We end up with brains specifically designed to pick up and copy religious memes. I suggest this is why religion, belief in God, and religious ritual still thrive in a modern scientific culture that rationally rejects them. Our brains are especially good at picking up these kinds of memes because of our long history of coevolution with them.

The argument takes the same form as before. Successful memes spread. They then change the environment in which genes are selected. The consequence is a brain that is better designed for spreading those particular memes.

The brain is good for genes as well as memes

In claiming that the human brain was built to copy memes, I have perhaps implied that it was of no benefit at all to genes (Blackmore, 1999). Yet our big brains have clearly provided all sorts of survival benefits enabling us to occupy widely varying niches all over the planet. My mistake was perhaps to overemphasize the role of the most arbitrary, useless, or even dangerous memes. But I did this mainly because it is on this issue that memetics diverges most strongly from traditional sociobiology or gene–culture coevolution theory.

In these disciplines genes provide the capacity for culture. Maladaptive (for genes) traits can arise and can even persist (Cavalli-Sforza and Feldman 1981; Feldman and Laland 1996) but competition between these traits is not taken into consideration, and benefit to the traits themselves is not the driving force. On the model I am proposing here, memes compete with memes and the outcome of that competition affects genes. Many memes survive precisely because they are useful to genes, but others survive for other reasons. They are not just maladaptive for genes; they are adaptive for themselves and for the memeplexes of which they are part. The whole package—the two-replicator creature—is an extremely efficient survival machine but we can only understand it by considering the effects of memetic competition. These effects are most obvious when they run counter to the interests of genes, so this is why I have tended to emphasize them.

I admit I had imagined the brain as having been driven to its huge size mostly by viral memes—in other words being somewhat analogous to a parasite that must be carried at some cost to genes—but this raises the interesting question of whether the brain is best seen as parasite, symbiont, commensal, or something else (a question raised in discussions with Pinker, Dennett, and some of their students). This has led me to the following analogy.

We can imagine the brain as analogous to the immune system. Memetic driving forces genes to produce a bigger brain especially good at copying any successful memes that are around. The genes fight back by producing ways of selecting only memes that are useful for them. This requires a complex system for recognizing which memes are useful and which not—something like the way the immune system has to recognize self from invader.

An example may help. Suppose that our hypothetical meme fountain is especially good at hunting with the latest tools, as well as dancing the latest flamboyant rain dance and flaunting his status by wearing the latest clothes. He has a survival advantage and so any genes he has that predispose him towards copying these memes are passed on. Other people copy him because he has the best memes, but there is another competition going on here. The people who selectively copy the useful tool skills from him while ignoring the dance will do better (biologically) than those who copy any and all of his memes. Although the simple heuristic—copy the meme fountain—works up to a point, the ability to select genetically beneficial memes from among those displayed by the meme fountain does even better. Meanwhile, memes are moving on in their purely memetic competition, outsmarting whatever selective tricks genes have come up with so far and putting on more pressure to be able to select among memes even more cleverly. The result is a brain that is very good at imitation, highly selective, and whose selective capabilities have been shaped by memetic competition.

Whether this comparison will prove useful I do not know. However, the central point here is this—on this theory, the brain was designed to copy successful memes, and this means both memes that succeed for purely memetic reasons, and memes that actually help survival of genes. In other words, it is a compromise between the forces of memetic and genetic evolution. Human intelligence is, in this view, all about the selection of memes, and future research should focus on which memes we do and do not copy, and why. This is a new way of looking at the function of human intelligence. The human brain is a selective imitation device.

Can memes get off the leash?

Lumsden and Wilson (1981) famously declared that 'the genes hold culture on a leash'. Most people who have modelled gene–culture coevolution have agreed. Even Durham (1991) who is one of the few to use the term 'meme', and who provides examples of maladaptive traits that spread successfully, argues that organic and cultural selection work on the same criterion—inclusive fitness. As far as I can see only Cloak (1975), and Boyd and Richerson (1985) truly treat their cultural trait as a replicator in its own right—an idea that is fundamental to memetics.

My arguments about meme–gene coevolution implied a complex interaction between the two replicators in which each affects the other—two dogs on the same leash you might say. But then the question arises whether the new dog can escape the leash altogether.

Among the factors that may be relevant are whether memes are transmitted vertically (parent to child) or horizontally (between unrelated people, possibly of the same age) (Cavalli-Sforza and Feldman 1981). A related issue is the relative speeds of change of the two replicators. If all memes are transmitted vertically then memetic change tracks genetic change and there can be no meaningful coevolution (indeed no leash at all). I have assumed that during most of human evolution memes were transmitted largely vertically, and changed at speeds not very different from human genetic change, but that there was sufficient horizontal transmission to make memetic driving possible. Nowadays, however, memetic transmission is very fast and largely horizontal. Although most people still get their first language, their basic social rules, and their religion from their parents, most of the memes they acquire during their lifetime come from school, radio, television, books, magazines, the Internet, their peers and even their own children.

In such an environment, genes can hardly be expected to track memes. They may still be affected, for example by memes such as birth control, technological medicine, genetic engineering, and so on, but memes are moving too fast for any detrimental effect they may have to exert any control. If the memes you come across are going to kill you or prevent you having children, the demise of your genes will come too late to exert any control over the spread of those memes. In other words, memes are getting off the leash.

Can this idea be formalised in any way? Bull has recently used an artificial life model to simulate the interactions between two replicators of different speeds (Bull *et al.* in press). When there is low dependence between the 'genes' and 'memes,' relative speed makes no difference to either replicator, but with slightly higher interdependence, increasing the rate of meme evolution provides rapid benefits to the memes, and gene evolution degrades to a random walk. This is only a simple and abstract model but suggests ways in which some implications of meme–gene coevolution might be tested.

Even if human genetic evolution is now no more than a random walk, it could still be argued that memes depend on genes for their propagation because they still build the brains which carry out the imitation—and it is these brains, with their endless penchant for food, sex, and violence, that determine the success of magazines, television programmes, and websites. In this sense, then, memes cannot be truly independent.

However, we may indulge in science fiction speculations and imagine the day—possibly not so far off—when humans are no longer required to maintain the hardware of the Internet because self-replicating computers have been designed. Even without that step, we can easily imagine information that is copied in the Internet without any human making the decisions. For example, there are already websites that generate a new academic paper, complete with references and footnotes, every time you visit. Imagine a program that chooses among these and distributes copies to other sites and you have memetic evolution without human intervention. Another possibility is that simple programs that currently pretend to be human users of chat rooms and discussion lists will evolve into much cleverer memeplexes, selectively copying behaviours from each other and from human users and so acting as autonomous, evolving, meme-selecting devices.

Such speculations are always dangerous but I mention this only to make a last, general point about the replicator power of memes. If memes are true replicators in their own right, as I have assumed, then we should expect them to coevolve along with the machinery for their own replication. Genes have done this—the exquisitely accurate machinery that copies DNA cannot have sprung into existence fully formed, but must have evolved gradually from simpler copying mechanisms (Maynard Smith and Szathmáry 1999). Memes are now doing the same. The process of meme–gene coevolution I have described can be seen as one step in this process—that is, the memes and the brains that copied them coevolved. But later steps are now far more significant. These include the invention of writing; the building of roads, railways and ships; the development of printing and books; the invention of the telephone, fax, and mobile phone; and most recently the Internet. Each step has improved the methods by which memes can be copied and stored, and made possible the creation of ever more memes. In our new memetic view of the universe, we should see these great steps in copying technology not as inventions consciously created for our own benefit, but as the inevitable consequences of memetic evolution. And *cui bono*? The memes. It is this process that may one day let memes right off the leash.

Conclusion

Memetics provides a new vision of human nature, in which memes succeed wherever and whenever they can. Memes spread not necessarily because they benefit either the genes that made their evolution possible, or the survival chances or happiness of the people who copy them, but because they benefit themselves.

In this vision, all the cultural entities around me are there because they are the current winners in a fearsome competition to be copied. My own body is a meme machine designed by a long history of meme–gene coevolution. It is furnished with plenty of memes it has already copied, and surrounded by masses more potentially copyable memes from which it has to choose.

On the optimistic side, there are several mechanisms by which altruistic behaviours may get themselves copied even though they are costly both to the person who carries them out and to their genes. Most simply, if altruistic people attract more friends who copy them then their altruistic behaviours obtain an advantage. Religions and cults may survive because they use clever memetic tricks to get themselves passed on, and to persuade their carriers to work hard and invest time and money in their propagation. Alternative therapies that do not work may thrive in modern environments, because of the powerful placebo effect combined with a fear of high-tech medicine. Even bizarre ideas like four-foot high aliens who come and abduct people from their beds at night can usefully be seen as memes that succeed despite being false.

Perhaps most challenging is the idea that my inner self, which seems to have consciousness and free will, is in fact a memeplex created by and for the replication of memes. 'My' beliefs and opinions are survival tricks used by memes for their own perpetuation. 'My' creativity is really design by memetic evolution. On this view, human nature is a product of memes and genes competing for replication in a complex environment, and there is no room for mysterious guiding principles or inner selves with free will.

In these and other ways memetics might utterly transform our view of ourselves. Indeed I suspect that taking the memes' eye view will provide as dramatic a transformation in our understanding of human nature as taking the genes' eye view has done in evolutionary biology.

References

Barton, R. A. and Dunbar, R. I. M. (1997). Evolution of the social brain. In *Machiavellian Intelligence: II. Extensions and Evaluations* (ed. A.Whiten and R.W.Byrne), pp. 240–263. Cambridge: Cambridge University Press.

Blackmore, S. J. (1999). *The meme machine.* Oxford: Oxford University Press.

Blackmore, S. J. (in press). Evolution and memes: The human brain as a selective imitation device. *Cybernetics and Systems.*

Boyd, R. and Richerson, P. J. (1985). *Culture and the evolutionary process.* Chicago: University of Chicago Press.

Brodie, R. (1996). *Virus of the mind: The new science of the meme.* Seattle: Integral Press.

Brugger, A. E. and Bushnell, E. W. (1999, April). Imitative strategies employed by 15- and 21-month old infants for learning to work novel objects. *Poster, Conference of the Society for Research in Child Development*, Albuquerque, NM.

Bull, L., Holland, O. and Blackmore, S. (in press). On meme–gene coevolution. *Artificial Life.*

Byrne, R. W. and Whiten, A. (ed.) (1988) *Machiavellian intelligence: Social expertise and the evolution of intellect in monkeys, apes and humans.* Oxford: Clarendon Press.

Campbell, D. T. (1960). Blind variation and selective retention in creative thought as in other knowledge processes. *Psychological Review,* **67,** 380–400.

Cavalli-Sforza, L. L. and Feldman, M. W. (1981). *Cultural transmission and evolution: A quantitative approach.* Princeton: Princeton University Press.

Cloak, F. T. (1975). Is a cultural ethology possible? *Human Ecology*, **3**: 161–82.

Dawkins, R. (1976). *The selfish gene.* Oxford: Oxford University Press.

Dawkins, R. (1993). Viruses of the mind. In *Dennett and his critics: demystifying mind* (ed. B. Dahlbohm), pp. 13–270. Oxford: Blackwell.

Deacon, T. (1997). *The symbolic species: the co-evolution of language and the human brain.* London: Penguin.

Dennett, D. (1995). *Darwin's dangerous idea.* London: Penguin.

Dennett, D. (1999). *The evolution of culture.* Charles Simonyi Lecture, Oxford, 17 February.

Donald, M (1991). *Origins of the modern mind: three stages in the evolution of culture and cognition.* Cambridge, MA: Harvard University Press.

Donald, M (1993). Precis of Origins of the Modern Mind: Three Stages in the Evolution of Culture and cognition. *Behavioral and Brain Sciences*, **16**, 737–91.

Dunbar, R. (1996). *Grooming, gossip and the evolution of language.* London: Faber & Faber.

Durham, W. H. (1991). *Coevolution: Genes, culture and human diversity.* Stanford: Stanford University Press.

Feldman, M. W. and Laland, K. N. (1996). Gene-culture coevolutionary theory. *Trends in Ecology and Evolution* 11, 453–7.

Gabora, L. (1997). The origin and evolution of culture and creativity. *Journal of Memetics—Evolutionary Models of Information Transmission 1.* [http: //www.cpm.mmu.ac.uk/jom-emit/1997/vol1/gabora_l.html]

Higgs, P. G. (in press). The Mimetic Transition: A simulation study of the evolution of learning by imitation. *Proceedings of the Royal Society.*

Jablonka, E. (1999, April). Between development and evolution: the generation and transmission of cultural variations. Paper presented at Conference on 'The Evolution of Cultural Entities', The British Academy, London.

Kendal, J. R. and Laland, K. N. (in press). Mathematical models for memetics. *Journal of Memetics–Evolutionary Models of Information Transmission.* [http: //www.cpm.mmu.ac.uk/jom-emit/]

Lumsden, C. J. and Wilson, E. O. (1981). *Genes, mind and culture.* Cambridge, MA: Harvard University Press.

Maynard Smith, J. and Szathmáry, E (1999). *The origins of life: from the birth of life to the origin of language,* Oxford: Oxford University Press.

Miller, G. F. (2000). The mating mind: *How sexual choice shaped the evolution of human nature.* London: Heinemann.

Pinker, S. (1997). *How the mind works.* London: Penguin.

Richerson, P. J. and Boyd, R. (1992). Cultural inheritance and evolutionary ecology. In *Evolutionary ecology and human behaviour* (ed. E.A. Smith and B.Winterhalder), Hawthorn, NY: Aldine de Gruyter. pp. 61–92.

Runciman, W. G. (1998). Greek hoplites, warrior culture, and indirect bias. *Journal of the Royal Anthropological Institute*, **4**: 731–51.

Tomasello, M., Kruger, A. C. and Ratner, H. H. (1993). Cultural Learning. *Behavioral and Brain Sciences*, **16**: 495–552.

Whiten, A. and Byrne, R. W. (1997). *Machiavellian intelligence: II. extensions and evaluations.* Cambridge: Cambridge University Press.

CHAPTER 3

Taking memetics seriously: Memetics will be what we make it

David L. Hull

Postmodern constructivists claim that nothing is really discovered. Everything is constructed, fabricated, or made. Although I am not all that fond of the relativist connotations of these terms, some things really are more made than discovered. For example, I do not think anyone ever discovered science. We constructed and reconstructed it through the years, and that construction process is far from over. Luckily, however, this process of construction is not totally open. At any one time, constraints on how we can construe science limit our freedom. Perhaps in the long run, any construct can be modified so that it is no longer recognizable, but in the short run the development of science is a matter of constrained construction.

For the past few years some young Turks have been urging a new science on us—the science of memetics. The goal is to treat conceptual change scientifically. But do not we already have a science that deals with conceptual change scientifically? It is called 'linguistics'. How then does this new science of memetics differ from linguistics, including quantitative linguistics? As far as I can tell, the only major difference between memetics and linguistics is that memetics is modeled on selection as it functions in evolutionary biology. Memetics then would be part of a more general research program designed to see which phenomena in addition to gene-based selection in biological evolution can be treated as selection processes; for example, the reaction of the immune system to antigens, operant learning, the development of the central nervous system, and possibly even conceptual change itself (Dawkins 1983; Cziko 1995). The last example is, of course, the most controversial.

Susan Blackmore (1999) proposes to carve out a specific niche for memetics in the midst of all the different entities that have been termed 'memes'. First, she distinguishes between individual learning (both classical and operant conditioning) and social learning. Individual learning might well count as a selection process (Glenn 1991), but it is not part of the subject matter of memetics because it cannot be passed from organism to organism in a copying process. For this same reason, immediate perceptions and emotions also do not count as memes. I feel my own pain. I can let others know that I am in pain in a variety of ways, but I cannot pass on copies of my pain to others. In order to count as a meme, the content (or form) of the meme must be passed on from organism to organism through imitation.

Blackmore (1999: 49) goes on to distinguish between social learning in general and one specific sort of social learning—imitation. Social learning in general involves observing others. As Heyes (1995) sees it, the difference between social learning in general and imitation turns on what is learned. 'Imitation is learning something about the form of behavior through observing others, while social learning is learning about the environment through observing others'. In any case, according to Blackmore (this volume), 'memes are, by definition, passed on by imitation'. As a result, imitation and hence memetics is limited almost exclusively to a fairly restricted sort of human behavior. All the usual examples of social learning, such as tits opening bottles of milk and Japanese monkeys washing sweet potatoes, turn out not to be instances of imitation and hence are not the concern of the new science of memetics (see Laland and Odling-Smee, this volume for the role of other forms of social learning in memetics).

I certainly can see the point of making the preceding distinctions, but limiting memetics to the study of imitation at the organismic level seems to narrow the subject matter of this science too drastically too soon. Blackmore (1999: 45 and this volume) argues that in individual learning 'nothing is copied from one individual to another, so there is no basis for a replicator to operate'. If we accept this argument, then we must conclude that the reaction of the immune system to antigens also does not count as selection because replication takes place at the cellular, not the organismic level. Single-celled organisms evolve through natural selection at the cellular level, but as soon as single cells gang up to form a multicellular organism, selection can no longer occur at the cellular level—but it does. The reaction of the immune system to antigens is as clear a case of selection as exists in nature.

Perhaps individual learning is not an instance of memetics, but not for the reasons that Blackmore gives. The issue here is not so much which processes count as selection processes, but which of these selection processes are the proper subject matter of memetics. For now, I would think that casting our net too broadly is a better strategy than casting it too narrowly, especially when the narrow definition of 'memetics' ends up with memetics being all but limited to human beings. One of the appeals of sociobiology, evolutionary psychology, and evolutionary epistemology is their apparent basis in evolutionary biology as such, not just in the evolution of one peculiar species.

Initially, evolutionary epistemology consisted of reasoning analogically from gene-based selection in biology to meme-based selection in conceptual change. This formulation of our research program opened the door to all the usual objections to analogical reasoning (e.g., disanalogies between genes and memes). Given an incredibly simplistic notion of genes, memes are not in the least like genes. The genetics that was incorporated into population genetics was Mendelian genetics. Mendelian genetics, so the critics claim, is particulate and concerns only pairs of alleles at a single locus (for a response, see Wilkins 1998a, b). In contrast, memes are not all that particulate, and more than two alternative memes can compete with each other. Of course, one does not need to know very much Mendelian genetics to know that Mendelian genes are not all that particulate and that numerous alternatives to Mendelian diploid inheritance exist (Crow 1979, 1999). One problem with interdisciplinary work is that any one worker is likely to know much more about one area than any of the others. Geneticists know much more about the complexities of genetics than of social groups. Conversely, anthropologists and sociologists tend to be well-versed in the details of social groups. To them genetics looks pretty simple. Contrary to what we were all taught in high school, genes are nothing like beads on a string. So both memes and genes are likely to have comparably complex structures.

But a more fundamental response to this objection is that memetics does not involve analogical reasoning at all. Instead, a general account of selection is being developed that applies equally to a variety of different sorts of differential replication. Instead of genetics forming the fundamental analog to which all other selection processes must be compared, all examples of selection processes are treated on a par. How well does each process accord to this general account of selection? If a feature of a particular example does not fit, does this mean that the example is not an instance of selection or must the analysis of selection be changed? One attempt at answering these questions can be found in Hull *et al.* (forthcoming).

Boyd and Richerson (this volume) argue that population thinking is more fundamental than natural selection in our conceptualization of culture in terms of material causes. The sort of variation that functions in selection is certainly necessary for selection. It is equally true that people have a very difficult time understanding, let alone accepting, the sort of variation that Mayr (1982) terms 'population thinking'. Just measuring a particular trait and finding its mean or mode is not very helpful in understanding the sort of variation that functions in the evolutionary process. In one part of the range of a species, one allele may be all but fixed. In another geographic location, another allele at that locus may be all but fixed. Averaging the two to produce a single population distribution would destroy the very information necessary to understand selection.

If species are taken to be the things that evolve primarily through natural selection, then convincing people that species have no 'essences' is far from easy. Even advocates of evolutionary psychology feel required to argue for a monomorphic mind—all people have essentially the same mind, a few deviants notwithstanding (Tooby and Cosmides 1990). Convincing people that sociocultural correlates lack an 'essence' is even more difficult. For example, Ernst Mayr (1983), the father of population thinking with respect to biological evolution, is convinced that evolutionary theory itself has an essence—a set of axioms that characterize the evolutionary process. But if conceptual systems are construed as evolving in anything like the way that biological species do, then they cannot be viewed in terms of eternal, immutable essences.

Conceptual clarity

Complaints about the lack of conceptual clarity in memetics arise in part because of an unreal view of how clear and uncomplicated certain familiar terms in science actually were or are. For example, look at the term 'gene' itself (Portin 1993; Blackmore 1999: 54). Was it all that clear when it was first introduced in 1909 by W. L. Johannsen (Wanscher 1975)? As was commonly done at the time, Johannsen declared that his gene concept was 'completely free of any hypothesis'. To some extent the Mendelian gene was operationally defined. The operations were embodied in Mendelian experiments. The name of the game was to discover patterns of inheritance and then posit the number and kind of genes that could produce that pattern. Of course, if any stretch of genetic material did not exhibit any variation, then neither genes nor alleles existed at these loci. In fact, these stretches of the genetic material could not even be termed 'loci'. Given this operational definition of Mendelian genes, much of the genetic material did not consist in genes. Only when a mutation introduced an allele did a Mendelian gene spring into existence.

As clear and operational as this gene concept was, no one utilized it all that consistently. Of course, lots of genetic material could not be subdivided into distinct genes by Mendelian mechanisms, but it was still genetic material and might one day be revealed by other mechanisms. Even limiting oneself to Mendelian breeding experiments, additional genetic units were discovered—mutons, codons, cistrons, and operons (Wilkins 1998a). With the advent of molecular biology, even more gene concepts were introduced—structural genes, regulatory genes, introns, exons, nucleotides, junk DNA, you name it. Mendelian geneticists complained that by terming all these molecularly defined entities 'genes', molecular biologists were destroying the clarity of the Mendelian gene. Of course, they did not mention that they themselves had already destroyed much of its clarity.

This scenario was replayed when G. C. Williams introduced his evolutionary gene concept. Just as Mendelian geneticists needed their gene concepts and molecular biologists needed their even larger array of genetic units, evolutionary biologists were justified in defining 'gene' for their needs. Even so, their critics (e.g., Stent 1980: 11) complained that evolutionary biologists were destroying the already well-understood molecular concepts. In his influential book, Williams (1966: 25) defined an evolutionary gene in terms of selection. An evolutionary gene is 'any hereditary information for which there is a favorable or unfavorable selection bias equal to several or many times its rate of endogenous change.' Dawkins (1976) adopted this definition and extended it to replicators in general. Since Dawkins is one of the fathers of memetics, one might continue this process and rework the Williams–Dawkins' definitions to apply to memes. According to Wilkins (1998a: 8; see also Wilkins 1999: 1):

A meme is the least unit of sociocultural information relative to a selection process that has favorable or unfavorable selection bias that exceeds its endogenous tendency to change.

I can already hear howls of derision. This definition is anything but operational! But why wait until this definition is extrapolated to memes to raise operational objections to it? Williams' definition of evolutionary genes is just as difficult to apply as is its memetic correlate. In general, critics of memetics assume standards so high for scientific knowledge that few, if any, areas of science can possible meet them

However, memeticists are not totally off the hook. The standard view among philosophers of science is that no theoretically significant term can be operationally defined (although their arguments seem not to have been totally successful; e.g., Gatherer 1998 and Marsden 1999; see response by Heylighen 1999). Just as Mendelian geneticists and molecular biologists have provided operational criteria for applying their gene concepts, so must advocates of memetics. These operational criteria will not be 'definitions' as philosophers use this term. At best, they will be highly context-dependent rules of thumb. Even so, if memetics is to be taken seriously, such criteria must be provided, and they cannot be provided from the seat of a comfortable rocking chair. They can emerge only as one sets about *doing* memetics. One of the messages of this chapter is that advocates of this new science should shift away from general discussions toward attempts to apply these terms to real cases (e.g., Pocklington and Best 1997).

But I don't know exactly what I am supposed to be doing. Until I get really clear about what a meme is, how can I conduct any empirical investigations on memes?

In this respect, memeticists are in the same position as any scientist working in a new area. You can not know that a particular sample is a sample of gold until you know what gold is, but you can not know what gold is without investigating particular samples of gold. But you can not know that a particular sample is a sample of gold. . . . The solution to this ineluctable circle is obvious if not very intuitively pleasing:

you work on all fronts simultaneously. Crude empirical investigations lead you to develop your theoretical perspective more clearly and extensively, and as it improves, you are in a better position to run more sophisticated empirical investigations, and so on. This process is better portrayed as a spiral than as a circle.

For example, I remember when I first came across Planck's Principle about how new theories do not triumph by convincing old scientists but by these old scientists dying off and their places being taken by young scientists who are better able to appreciate these new ideas. I thought I knew what Planck meant, and since I was young at the time, I agreed with him. Even though I was sure that Planck's principle was correct, I decided to test it anyway. Is there any correlation between the age of scientists and how quickly they accept new scientific ideas? To say the least, attempting to test this apparently straightforward claim was a learning experience. Who counts as a scientist? What does it mean for scientists to reject or accept a new idea? How much of a theory must they accept before they can be considered as accepting it? What makes it 'new'? Very similar ideas are raised over and over again. For example, how 'new' is the meme concept? I would never have realized how serious these problems are if I had not tried to test Planck's Principle. As it turns out, age does not explain very much of the variation in the acceptance of new scientific ideas, at least not in the case of species evolving (Hull *et al.*, 1978).

The primary message of this chapter, then is that memeticists cannot begin to understand what the science of memetics is until they generate some general beliefs about conceptual change and try to test them. These tests are likely to look fairly paltry, but in the early stages of a science, attempts at testing always look fairly paltry. For years, just about the only example we had of the effect of selection on species was the peppered moths in industrial England. As it turns out, in retrospect, these studies were seriously flawed (Majerus 1998). Perhaps they convinced people at the time, but on closer inspection, perhaps they should not have. I want to urge memeticists to ignore the in-principle objections that have been raised to memetics no matter how cogent they may turn out to be and proceed to develop their theory in the context of attempts to test it. Continued semi-popular discussions of memetics are likely to have the same effect on 'meme' as they have had on 'paradigm' (Wilkins 1998a: 2). Quick and easy metaphors and popular science are likely to lead to the 'debasement of memetics' (Wilkins 1999: 6).

Memetics as a progressive research program

Robert Aunger, in his Introduction, sounds an alarm with respect to the very strategy that I am urging. If memetics were really a new research program, perhaps rough-and-ready attempts to test it would be justified, but the science of memetics has been around for over twenty years without exhibiting any significant conceptual or empirical advances. It is about time that we realize that it is clearly not a progressive research program. I agree with Aunger's evaluating emerging research programs on the criteria suggested by Lakatos (1970). However, Aunger's conclusion follows only if his dating of the origin of memetics is accurate. For Aunger, Dawkins' *The Selfish Gene* (1976) inaugurated the science of memetics. After all, he coined the term 'meme'. Other authors such as Blackmore (1999: 219) also date the origin of memetics with Dawkins' coining the term 'meme' in 1976.

But authors had been urging the studying of conceptual and cultural change as a selection process long before Dawkins, and several of these early authors coined new terms to refer to the units of this evolution. For example, in 1904 Richard Semon published a book entitled *Die Mneme als erhaltendes Prinzip in Wechsel des organischen Geschehens*. He published an English translation of this book in 1914 entitled *The Mneme*. Why not date the beginnings of memetics (or mnemetics) as 1904 or at the very least 1914? If these two publications are taken as the beginnings of memetics, then the development of memetics has been even less progressive than Aunger claims. It has been around for almost a hundred years without much in the way of conceptual or empirical advance!

But, once again, such a conclusion depends on the proper dating of memetics. How important are neologisms or first publications in the dating of scientific research programs? Is terming certain scientists 'unappreciated precursors' so wrong-headed? After all Mendel published his famous paper in 1865, but nothing much happened until the turn of the century when Mendelian genetics really took off. Should we date Mendelian genetics from 1865 or 1900? In 1964, W. D. Hamilton published his paper, 'The genetical evolution of social behavior', but it lay fallow for a dozen years or so. G.C. Williams' 1966 book took at least as long before other evolutionary biologists began to take it seriously. For purposes of determining progress, should we date research programs from their first hint in a publication or from the time at which scientists begin serious investigation? I think the latter is more appropriate. Not until a reasonable number of scientists begin to work on a new research program can its progress or lack of progress be measured.

The preceding problem is itself a major concern in the science of memetics. How do we decide when a 'new' idea was introduced? Is memetics fairly new, twenty years old, or a hundred years old? In his critique of memetic models, Best (1998) traces evolutionary models of cultural evolution from before Darwin, through the early 1970s to the present. Best is one of the most recent members of a research program that has a long history. But if we take memetics seriously, unappreciated precursors do not count. Nor do long-dead research programs. Although Semon had at least some impact in his own day, his views have had no impact on present-day advocates of memetics. As a recent exchange over who really coined the term 'meme' indicates, Semon is yet another unappreciated precursor (Laurent 1999). It also indicates that some people take the introduction of neologisms seriously, as if scientists did not exist until William Whewell introduced the term 'scientist' in 1834 and immediately rejected it as totally inappropriate. Scientists came into existence when Whewell rejected the term 'scientist'? The fascination with neologisms that so mesmerizes postmodernists continues to mystify me.

Just as 1900 seems the appropriate date for estimating the progressive character of Mendelian genetics, memetics should be evaluated only when a reasonable number of people began to develop it. The appearance of Dawkins' *The Selfish Gene* in 1976 did inaugurate a vast literature on *gene selectionism*. Picking 1976 as the beginning of this research program and judging how progressive it has been is accordingly justified. But Dawkins' suggestion about *memes* did not exactly take off in 1976. Several authors published books on the general topic of cultural evolution during the past twenty-five years or so; for example, Lumsden and Wilson (1981), Cavalli-Sforza and Feldman (1981), Boyd and Richerson (1985), Hull (1988a), Barkow (1989), and Durham (1991). All of these publications have their merits, but what all of them failed to do is to initiate an active research program in something that might rightly be termed 'memetics'.

How can we tell? One way is to engage in a bit of memetics: do citation analyses to see if one or more of these hopeful founders succeeded. My intuitive guess is that memetics as an active research program is quite new, no older than a dozen years. During this period numerous workers from a variety of backgrounds have devoted themselves to expanding on the notion of memetic evolution—and no standard higher than voting with one's career exists in science. Just as evolutionary biologists started the journal *Nature* as an outlet for their works and cladists began *Cladistics*, memeticists now have their own *Journal of Memetics*. That the British Academy sponsored a conference on memetics in April 1999 and that another conference was held at Cambridge that same year are additional signs that memetics has emerged as an active research program. How progressive it will turn out to be is another matter, but to make such judgments, we need to date the beginning of this research program appropriately.

Now that the science of memetics has begun to develop, the clock is ticking. Progress must be forthcoming. As I see it, the two areas in memetics most ripe for progress are the reconstruction of conceptual phylogenies and improving our understanding of the mechanisms involved in memetic transmission. In a previous publication (Hull 1995), I detailed the near identity of the methods used by paleontologists and biological systematists on the one hand, and historical linguists on the other in producing their respective phylogenies. Independently they have devised the same method of representing phylogenetic relations—the cladogram. Although cladograms are presented differently in the two disciplines (one with its peak pointing up, the other pointing down), they are designed to represent exactly the same relations. Workers in the two disciplines discovered the same problems and presented the same array of solutions. For example, both groups were forced to recognize the difficulties implicit in the comparative method with respect to distinguishing ancestral languages and ancestral taxa, respectively. Because I have presented my views on this subject elsewhere, I will not discuss this issue here other than to note that reconstructing linguistic phylogenies is as progressive a research program as is its correlate in biological systematics (e.g., Hoenigswald and Wiener 1987; Diamond 1988; Barbrook *et al.*, 1998; Croft 2000).

A persistent bias towards genes and organisms

One of the chief obstacles in understanding memetic evolution as a process is the hold that genes and organisms have on all of us. Dawkins

(1976) and I (Hull 1988b) suggested more general conceptions for understanding selection processes. Dawkins contrasted replicators with vehicles. The relation that he specified between these two classes of entities is development. Replicators produce, code for, ride around in, and direct vehicles. The influence of genes and organisms on Dawkins' conception is obvious. Just as the relation between genes and organisms is developmental, so is the relation between replicators and their vehicles. Hull (1988b) agrees with Dawkins' treatment of replicators but suggests an alternative to vehicles that is not limited to development—interactors. Development is a common, though not universal, mechanism for relating replicators and interactors. Any entity that interacts with its environment in ways that make replication differential is an interactor. Which causal relations produce this correlation is an open question, and development is not the only answer.

Although the notions of vehicle and interactor may seem quite similar, they differ in certain important respects. On my view, genes can function as both replicators and interactors. Quite obviously, genes can function as replicators, but they also interact with their cellular environments. They have adaptations (e.g., they are structured to replicate). Although replication is largely concentrated at the level of the genetic material, environmental interaction occurs at a variety of levels, from genes and cells to organisms and hives and possibly even demes and entire species. The relation between genes and lower level interactors can well be developmental, but as the interactors become more inclusive, the effects of development decrease. As has become clear in the literature, the levels of selection controversy concerns the level (or levels) at which environmental interaction is taking place, not replication.

Another bias that is introduced by the gene–organism perspective is setting out general accounts of selection in terms of *entities*. Genes and organisms are entities. Hence, the best way to characterize the selection process is in terms of more general entities—replicators and vehicles (or interactors). However, selection is a process. Hence, it might be better to explicate this notion in terms of processes, not entities. Selection is a process by which environmental interaction produces differential perpetuation. Dawkins (1982) already made a step in this direction by extending the phenotype beyond organismal limits. Traits tend to come bundled into organisms, but they need not. Treating selection and its two subprocesses as processes helps to circumvent a variety of problems. For example, Ghiselin (1999: 15) gets considerable enjoyment in pointing out that chromosomal deletions count as Dawkinsonian replicators because they can be favored by selection. The absence of a segment of DNA can count as a replicator (see Dawkins 1982: 164)?

Boyd and Richerson (this volume) point out that cumulative adaptive evolution is possible in the absence of replication or replicators. All that is needed is heritable variation. Mechanisms that do not involve modification through descent could serve the function that descent does in selection. However, thus far descent is the only mechanism that has evolved to produce the necessary correlations. A more general analysis might be couched in more abstract terms, but I happen to be strongly inclined toward mechanisms. In addition to abstract correlations, I want to know how the system works. Any adequate understanding of selection, as I see it, requires the specification of the mechanisms that are bringing about these correlations, even if other mechanisms are possible.

One important difference between Dawkins' analysis of selection and mine is that Dawkins (1994) introduced his notion of vehicle only to bury it. I argue that environmental interaction is a necessary part of the selection process. It is present at a variety of levels of organization and cannot be eliminated without serious explanatory loss. Anyone who wants to understand the mechanisms that are operative in a certain instance of selection must refer both to replication and to the relevant environmental interactions. Dawkins is a gene replicationist. Genes are the primary replicators in biological evolution. He can be counted a genetic selectionist only if he thinks that environmental interaction is irrelevant to selection processes. He does not. After all, replication without environmental interaction is, by definition, drift, and he thinks that there is more to biological evolution than drift.

Dawkins is also a reductionist of sorts. He thinks that, in the last analysis, fitness at higher levels of organization can always be reduced to fitness at the level of genes. Certainly, most population geneticists reason in this way when they are engaged in their professional studies. When they step back from their work and reflect on it, some retreat from the reductionist position is implicit in their own research. Others happily admit to being reductionists. I have only two contributions to make with respect to this eternal philosophical dispute. First, the general analysis that one proposes for selection is independent of one's position with respect to reduction. I think that both replication and environmental interaction are necessary for selection. I may or may not think that environmental interactions occurring at higher levels can be reduced to replication at the lowest level possible. In addition, 'reduction' does not entail 'reducing away'. All the entities that play causal roles in selection remain part of the selection process, regardless of the success or failure of reduction.

As strange as it may seem, the tendency of thinking in terms of genes and organisms pervades the literature on memetic evolution and gives rise to numerous misunderstandings. For example, one commonly hears that conceptual evolution is so much faster than gene-based biological evolution. Certainly, memes can be transmitted much more rapidly than the genes of such organisms as whales, people, and sequoia trees. However, even from the organismal perspective, viruses and bacteria reproduce themselves much more rapidly than the vast majority of memes. In this respect, I see no significant difference between genes and memes. Some genes are passed on quite rapidly; some quite slowly. Some memes are also passed on quite rapidly; others—sad to say—get passed on only very slowly. Darwin published his theory of evolution in 1859. A century and a half later, the vast majority of human beings have never heard of Darwinian evolution. Of those who have heard of it, the vast majority do not understand it. Of those few who do understand it, most do not accept it. This is speed? Only by ignoring all those organisms that reproduce with extreme rapidity as well all those memes that propagate at an incredibly slow pace can memetic evolution be made to look all that much faster than genetic evolution.

A second instance of the deleterious effects that the gene–organism perspective has had on memetics can be found in the frequency with which conceptual change is termed 'Lamarckian'. One topic that memeticists might investigate is the near compulsion that people writing on evolution have to finding some phenomenon to label 'Lamarckian'. In its literal sense, inheritance is Lamarckian if the environment changes the phenotype of an organism in such a way that this organism is better adapted to the environmental factor that produced this change. This phenotypic change must then be transmitted somehow to the genetic material so that it can be passed on to the offspring of the organism through reproduction. These offspring then are born with this acquired characteristic more highly developed or with a strong tendency to produce this characteristic more highly developed. Lamarckian inheritance is the literal inheritance of acquired characteristics. The transmission must be genetic, and the relevant effect must be phenotypic. For example, a mother dog might give her puppies fleas, but this transmission is not Lamarckian because it is not via genes. In addition, one might have some reservations about terming an organism's parasites part of its phenotype, but on this score I am willing to go along with Dawkins' attempt to 'extend' the phenotype.

In memetic evolution, new memes are certainly acquired. For example, you did not understand the Pythagorean theorem. You take a course in plane geometry, and now you do. You have acquired a new meme. You in turn can pass this increased knowledge on to someone else. Isn't this an instance of the inheritance of acquired characteristics? Not in the least. In the science of memetics, memes are analogous to *genes*, not *phenotypic characteristics*. Hence, if memetics is anything, it is the inheritance of *acquired memes*. How passing on memes (or fleas for that matter) can count as Lamarckian inheritance in any comprehensible fashion continues to elude me. It is made to look plausible only by running together the genetic with the memetic perspectives.

In gene-based biological evolution, memes might well be viewed as characteristics. Some of these memes may well have some genetic basis. In a literal sense they might be passed on genetically, but in no case to my knowledge has Lamarckian inheritance played a role. In meme-based conceptual or sociocultural evolution, memes are viewed as analogous to genes. Hence, no matter what they might or might not do, the result cannot be the inheritance of acquired characteristics. Enough said? I doubt it. Just as people insist on believing that female praying mantises eat their mates during copulation, beginning at the head so as not to interfere with the mating process, that perfectly good tasting viceroy butterflies avoid predation by mimicking the appearance of the nasty monarch butterfly, that lemmings periodically rush to the sea in order to commit mass suicide, that Darwin's finches played a crucial role in the development of his theory of evolution, and that Karl Marx wrote to Darwin asking to dedicate *Das Kapital* to this reclusive biologist, I am sure that the compulsion to term memetic evolution 'Lamarckian' will not be diminished by any argument anyone might present. Conceptual selection does not guarantee truth. In one's more cynical moods, one might complain that it almost never does.

The distinction between Lamarckian and non-Lamarckian inheritance turns on the genotype-phenotype distinction. One reason why conceptual change tends to look deceptively Lamarckian is that this distinction is not easy to make in memetic change. Memes play the role of genes in replication, but what counts as environmental interaction? In gene-based selection in biological evolution, vertical transmission is commonly distinguished from horizontal transmission. In vertical transmission, genes are passed down from parents to their offspring regardless of whether the form of inheritance is sexual or asexual. Any other form of transmission is horizontal. In biological evolution the only form of genetic transfer that seems in the least horizontal is infection by viruses. A virus can pass from one organism to another in two ways: during the reproduction of its host and independent of that reproduction. In the first instance, this transmission might be termed 'vertical'. After all, it is proceeding from parent to offspring. When it is transmitted to any other organisms, including organisms that belong to different species, this transmission is horizontal. However, all of the preceding comments are made from the perspective of the host, not the virus. From the perspective of the virus, all of its genetic transmission is vertical, and with respect to the fitness of the virus, this is the perspective that counts.

Nearly everyone who discusses memetic transmission claims that it can be both vertical and horizontal. If parents teach their offspring something, that is vertical. Any memetic transmission that differs from this genealogical direction is horizontal. The preceding claims follow, however, only from the perspective of organisms and their genes, but this is not the appropriate perspective for memetics. The basic entities in memetic evolution are memes, and their replication determines the direction of transmission. One might well emphasize when memetic transmission differs from genetic transmission, but the distinction between vertical and horizontal transmission that is relevant in memetic evolution must be in terms of memes, not genes.

In old fashioned evolutionary epistemology, such things as behavior are treated as *characteristics* partially under the control of genes (e.g., the genes that promote sucking in newborn mammals). But in modern memetics, memes are analogous to genes, not characteristics. If genes determine which transmissions are vertical in traditional gene-based selection, then memes must determine which transmissions are vertical in meme-based selection. The reason that this conclusion is unsettling is that we do not have a very clear idea of memetic environmental interaction. While memetic replication seems clear enough, memetic environmental interaction does not.

Replication and its implementation

One of the most crucial but difficult tasks that confronts memeticists is formulating in the context of conceptual change something analogous to the distinction between genes and phenotypic characters. Gabora (1997) discusses this distinction in terms of information and its 'implementation'. However, Gabora limits memes to mental representations and treats their implementation in behavior or artefacts as the phenotypes of these mental representations. I think that this way of dividing up the subject matter of memetics is mistaken. As long as information is passed along largely unchanged, the process counts as replication regardless of the substrate ('vehicle' in Campbell's sense). The printed page, a floppy disk, a magnetic tape, the spoken word, sign language, and even vibrations in the air are all capable of incorporating information in their make-up and transmitting it via replication. We do not know enough about the brain yet, but it seems very likely that brains can also contain and transmit information (Baddeley and Hancock 1999).

However, certain memeticists recoil from treating 'unobservable' mental entities as replicators (Gatherer 1998; Marsden 1999) even though generations of philosophers have repeatedly refuted the operational philosophy on which such rejections are based. Even Skinnerian behaviorists have overcome their blanket rejection of mental entities. If I have any advice for memeticists, it is not to embrace the most infamous of all philosophical tar babies—the mind/body problem. All I can say in this respect is that perhaps 'phenomenal givens' cannot enter into the causal sequences that produce replication, but for every phenomenal given, there has to be something going on in the brain, and these neuronal memes will do just as well (see comments by Speel 1999). Any deeper probing into this tangle of problems is guaranteed to detour any career for a lifetime. Conceptual change involves the pooling of conceptual resources, and any memeticists who want to get on with their research program can cite, without apologies, the work of Dan Dennett (1991, 1995) and then move on. That is what philosophers are for.

Such philosophical warnings to one side, we still must find some way to distinguish between replication and its implementation. The idea that seems central is *information* (Maynard Smith and Szathmary 1995). Lake (1998) presents a promising discussion of memetic 'phenotypes' in terms of decoding. For Lake (1998: 82):

... replicators are information, that is to say, they are symbolic structures which code for, or refer to, non-symbolic structures. If a replicator passes on its structure directly then replication must be a process in which symbolic structure is transmitted without decoding. For sure, the symbolic structure often is decoded, but it is part of the process of interaction, not replication. In the case of biological evolution, for instance, genes provide information about how to build an organism. The fitness of the organism determines the frequency with which the genes that coded for it are replicated, but these genes are never re-encoded. . . .

Fleshing out the notion of 'decoding' is more difficult than one might think. The notion of 're-encoding' is just as difficult. Two issues, both highly problematic, converge in making the relevant distinctions so difficult to get a hold of. For one thing, we do not have a very clear notion of what information is, at least not a notion of information up to the tasks required of it in selection processes. According to thermodynamicists, all structures have or contain information. The solar system, an enclosed gas, and a molecule of table salt all contain information. So does a molecule of DNA. It is a double helix. The bonds that run along the 'backbones' of this molecule do not rupture as easily as those holding the base pairs together. Hence, the molecule can zip and unzip with great facility. However, another sort of information is also contained in a molecule of DNA—in the sequence of its base pairs. As far as I know, none of the current analyses of evidence can distinguish between these two sorts of information, and until they do, memetics is in real trouble. The people working in information theory cannot distinguish between the information contained in the structure of the paper on which this book is printed and the information contained in the printed sequence of letters and words. That they cannot is a scandal. The problem is ripe for solution.

A second even messier problem turns on the asymmetry between the ease of reading the information contained in a meme into some application and the difficulty of the opposite inference. Copying instructions is relatively easy. Inferring the instructions from the product is extremely difficult. Complicating this asymmetry is the old nature–nurture issue. As Wilkins (1998a: 13) remarks, genes do not code for traits but for reaction norms. Given clones of a single genotype, the resulting organisms can vary tremendously depending on variations in the environment. The relation between a genotype and its possible phenotypes is one-to-many. Conversely, given a single phenotypic characteristic, numerous possible combinations of genes and environmental variables could have produced it. The net result is that development is all too frequently a many-to-many relation (Wilkins 1999: 3).

However, the asymmetry between making a product using instructions and inferring these instructions from the product is different from the familiar nature–nurture issue. Dawkins (1982) explicates this relation in terms of a cake and its recipe. If one already has the skills necessary to bake cakes, it does not take much to bake numerous, almost identical cakes from a recipe. Although the resulting cakes may vary somewhat because of such things as differences in altitude, different sizes of eggs, and outright mistakes, in general the relation between a recipe and a cake is close to one-to-one. One could also learn how to bake a cake by watching someone else bake a series of cakes even in the absence of a recipe. (Of course, the ideal state is to have a recipe and watch the implementation of this recipe.) However, reconstructing a recipe from a cake itself is more difficult. Too many different recipes and too many alternative skills could have been used to bake this cake. The relation is hardly one-to-one.

Blackmore (1999: 51, 214) characterizes this difference as an example of reverse engineering. Copying instructions for making a compact disk player is easy. Given these instructions and some general technological knowledge, making the disk player is also relatively easy. But unscrupulous manufacturers try to circumvent patents by attempting to copy the product itself, that is, they try to infer the instructions for making the product from the product—a much more difficult undertaking. Boyd and Richerson (this volume) illustrate these same points in the context of making a clay bowl. The three relevant elements are the written instructions for making clay bowls of this type, watching someone make such bowls, and the clay bowl itself. Boyd and Richerson conclude that memes are not much like genes because in conceptual evolution, for 'any phenotypic performance there are potentially an infinite number of rules that would generate that performance'. Although I think that 'potentially infinite' is a bit of an exaggeration, genes and memes do not differ all that much in this respect. For any phenotypic character, an extremely large number of genotypes would generate that character.

The purpose of the preceding discussion is to show how a series of replications can be distinguished from the translating of the information contained in these replicators to make a product—homocatalysis versus heterocatalysis. In this decoding process, a tremendous amount of information is lost. As a result, the product can at most function as a replicator with a seriously reduced information content. In general, it cannot function as a replicator at all. In sum, if we are to make sense of memetic evolution, we must free ourselves from the hold that genes and organisms have on us. The more appropriate and general terms are 'replication' and 'interaction,' and these relations can be distinguished via the transmission versus loss of information—if only we had a better understanding of what information actually is.

The memetic process

In the early pages of this chapter I recommended that advocates of memetics turn more of their attention from general discussions of memetics (such as this one) to testing elements in this research program. Hence, I am under some obligations to follow my own recommendations. Given our general understanding of memetics, what should we expect it to be like?

One of the perennial problems in biological evolution is that it proceeds much more quickly than the mechanisms available imply that it should. One solution to this problem is emphasizing the role of small groups of organisms, in particular, peripheral isolates. Once a species is sufficiently well established, sheer numbers make change excruciatingly slow. However, change in small isolated populations can be much more rapid because the demise of a very few organisms can make a significant difference. Not only selection but also drift can result in fairly rapid change. But small population size is also likely to lead to extinction because of increased inbreeding and the resulting homozygosity. Lande (1988) in turn has argued that random demographic and environmental events drive small populations to extinction before such genetic factors can come into play. Recently, several workers tested this hypothesis for butterfly populations in southwestern Finland and found significant effects of inbreeding (Saccheri *et al.*, 1998). This example also nicely illustrates an important feature of testing for memetic selection: it is not going to be easy.

Do comparable observations hold for memetic evolution? Listing a dozen or so cases of rapid conceptual change is easy. How many of them are associated with a relatively small, isolated community, as distinct from a single individual or large, unstructured communities?

In her study of high energy particle physics, Blau (1978) discovered that roughly half of these scientists were organized into small research teams, while the rest worked primarily alone. About half of these research teams were organized into a large invisible college. As it turns out, the scientists working in small research teams that were part of this larger invisible college were much more productive than either isolated scientists or isolated research teams. I found the same correlations with respect to the research groups that I studied (Hull 1988a).

Another distinction that is important in biological evolution is inter-specific versus intraspecific competition. In *Science as a Process* (Hull 1980) I studied two research groups—the numerical taxonomists originally located at the University of Kansas and then at the State University of New York at Stony Brook, and the cladists who gained their first toehold at the American Museum of Natural History. Did members of these communities treat fellow members of their group differently from scientists working in the other research group? If science is a selection process, one would certainly expect so. To test this hypothesis I studied all the manuscripts submitted to *Systematic Zoology*, the primary journal in this field at the time, and all the referee reports obtained for these papers for a period of seven years. Did cladists referees treat papers submitted by fellow cladists more gently than papers submitted by non-cladists? To my dismay, I could find no such correlations. Cladists were just as hard on cladists as they were on non cladists.

In the best tradition of science, I did not immediately reject my hypothesis but put it in the bottom drawer. Eventually I discovered what was going on. During the period that I was studying, cladists were beginning to speciate into transformed cladists and phylogenetic cladists. Prior to anyone noticing that this speciation was occurring, my study of refereeing patterns picked it up. When I went back to my data and distinguished between those cladists who eventually became transformed cladists and those who came to owe their allegiance to phylogenetic cladistics, the pattern that I had expected became clear. I had transformed an apparent falsifier into a confirming instance, one of the strongest indicators that a research program is progressive.[1]

---

[1] I also discovered a second error that I had committed in my original study. I had divided the systematists whom I was studying into cladists and non-cladists, a habit that cladists had through the years convinced me is a serious mistake. We should not divide animals into vertebrates and invertebrates. Invertebrates is a waste basket taxon. So is non-cladist. I should have compared cladists with numerical taxonomists.

One final example of similar processes operating in biological and memetic change is kin selection. Organisms tend to treat close relatives differently from other organisms in their species. In kin selection genealogy matters. Of course, the notion of genealogy must be operationalized. One operationalization is in terms of the organisms that one bumps into first. Proximity in early development operationally defines 'kin'. Mistakes will be made, but that is only to be expected. The immune system seems to use this same method to distinguish between self and non-self. In science, scientists also distinguish between kin and non-kin, but the relevant genealogy is conceptual. The issue is not who holds similar ideas but who is conceptually connected to whom. The best way to increase the likelihood that you will be a successful scientist is to work under a successful scientist (Hull 1988).

One sort of experimental study that memeticists have undertaken is to trace the differential perpetuation of memes through time. The Internet supplies a mother lode of data to be mined. For example, Pocklington and Best (1997) traced such cultural replicators as 'Nazi' on the Internet in a particular linguistic pool to ascertain patterns of relative transmission.[2] As another example, Dawkins himself (1999: viii) records the number of times that 'memetics' is mentioned on the World Wide Web. As of 12 August 1998, it had been mentioned 5042 times compared to the 'extended phenotype' (515) and 'exaptation' (307). The implications are obvious. Memetics is proving much more successful than Dawkins' own extended phenotype and Gould and Vrba's (1986) exaptation—at least on the Web. I fear to do a similar count for 'interactor'.

As in the case of biological evolution, numbers are not enough. Biologists want to know more. They want to know what is *causing* these changes. Professionals are much more interested in professional versus popular uses of their terms. *Nature* and *TV Guide* are not of equal worth when it comes to estimating the impact of an evolutionary biologist's views about evolution on his or her fellow evolutionary biologists. It is also important whether the term is used as a substantive part of an author's own work or is introduced only to be dismissed. Acceptance is best, but dismissal is better than no mention at all. Academics are justifiably suspicious of citation analyses if they do not recognize the

---

[2] Pocklington and Best (1997) used principal-components analysis to discover their patterns of transmission. Using a cladistic algorithm, such as PAUP, would have produced a more accurate picture, but sheer numbers precluded such studies. Too much computer time would be required.

preceding distinctions. One author accumulates an increasing number of citations because later authors incorporate his work into their own. Another author accumulates an equally impressive list of citations but because later authors reject his views. The numbers may be the same. The causes and their implications are significantly different. Discovering what is causing changes in meme frequencies is likely to prove as difficult as determining the causes of changes in gene frequencies.

Such challenges are difficult but not impossible to meet. For example, in a couple of hundred years a population memeticist might notice a strange transition in English over the past couple of decades. The frequency of the use of 'he' has dropped while the occurrence of such barbarisms as 'he/she' and 's/he' increased quite markedly. More dramatic still, 'stewardess' has disappeared to be replaced by 'flight attendant', 'mailman' has been replaced by 'mail carrier', and now your meals are brought to your table by a member of the 'wait staff'. Literally hundreds of similar changes have occurred in a short period of time. Might it be possible that a future memeticist tumble to what is going on and works out why?

Conclusion

Memetics is an emerging research program like any other. It should be evaluated the way that other research programs are evaluated. Is it progressive? During the past decade or so, I think that it has shown considerable progress, but to succeed it will have to continue on this trajectory. Increased coherence and articulation is certainly worthwhile, but such improvements can occur only in conjunction with attempts at testing. Testing is not easy but it is necessary. Eventually, advocates of memetics will have to respond to the fundamental objections that its opponents are raising to it. Some of these objections may turn out to be met without significant reworking of this emerging research program. Others may require extensive reformulation. Still other objections may simply be misguided. But for now, memeticists need to close ranks and work together to develop their program. Science is an extremely high resolution activity. Very few new research programs ever gain much currency. Even fewer succeed, but in such matters the results are worth all the effort. What if we do develop a theory of population memetics that can do for conceptual and sociocultural change what traditional population genetics does for biological evolution? That would certainly be worth the effort.

Acknowledgements

I wish to thank Robert Aunger for an extensive discussion of the issues raised in this paper.

References

Baddeley, R. and Hancock, P. (1999). *Information theory and the brain*. Cambridge: Cambridge University Press.

Barbrook, A. C., Howe, C. J. Blake, N. and Robinson, P. (1998). The phylogeny of 'The Canterbury Tales'. *Nature*, 394: 839.

Barkow, J. H. (1989). *Darwin, sex and status: Biological approaches to mind and culture*, Toronto: University of Toronto Press.

Best, M. L. (1998). Memes on memes: A critique of memetic models. *Journal of Memetics–Evolutinary Models of Information Transmission*, 2. [http: //www.cpm. mmu.ac.uk/jom-emit/]

Blackmore, S. (1999). *The meme machine*. Oxford: Oxford University Pres,.

Blau, J. R. (1978). Sociometric structure of a scientific discipline. *Research in Sociology of Knowledge, Sciences and Art*, 1: 191–206.

Boyd, R. and Richerson, P. J. (1985). *Culture and the evolutionary process*. Chicago: University of Chicago Press.

Cavalli-Sforza, L. L., and Feldman, M. W. (1981). *Cultural transmission and evolution: A quantitative approach*. Princeton: Princeton University Press.

Croft, W. (2000) *Explaining Language Change: An evolutionary approach*. London: Longman.

Crow, J. F. (1979). Genes that violate Mendel's rules. *Science*, 240: 134–46.

Crow, J. F. (1999). Unmasking a cheating gene. *Science*, 283: 1651–52.

Cziko, G. (1995). *Without miracles: universal selection theory and the second Darwinian revolution*. Cambridge, MA: MIT Press.

Dawkins, R. (1976). *The selfish gene*. Oxford: Oxford University Press.

Dawkins, R. (1982). *The extended phenotype*. Oxford: Oxford University Press.

Dawkins, R. (1983). Universal Darwinism. In *Evolution from molecules to men* (ed. D. S. Bendall), pp. 403–25. Cambridge: Cambridge University Press.

Dawkins, R. (1994). Burying the vehicle. *Behavioral and Brain Sciences*, 17: 616–17.

Dawkins, R. (1999). Foreword to *The meme machine* by Susan Blackmore, Oxford: Oxford University Press.

Dennett, D. (1991). *Consciousness explained*. Boston: Little Brown.

Dennett, D. (1995). *Darwin's dangerous idea*. London: Penguin.

Durham, W. H. (1991). *Coevolution: Genes, culture and human diversity*. Stanford:Stanford University Press.

Diamond, J. M. (1988). Genes and the Tower of Babel. *Nature*, 336: 622–3.

Gabora, L. (1997). The origin and evolution of culture and creativity. *Journal of Memetics–Evolutionary Models of Information Transmission*, 1. [http: //www.cpm. mmu.ac.uk/jom-emit/vol/gabora_1.html]

Gatherer, D. (1998). The case for commentary. *Journal of Memetics–Evolutionary Models of Information Transmission*, 3. [http: //www.cpm.mmu.ac.uk/jom-emit/]

Ghiselin, M. (1999). Darwinian monism: The economy of nature. In *Sociobiology and Bioeconomics* (ed. P. Koslowski), pp. 7–24. Berlin: Springer-Verlag.

Glenn, S. S. (1991). Contingencies and metacontingencies: Relations among behavioral, cultural, and biological evolution. In *Behavioral analysis of societies and cultural practices* (ed. P. A. Lamal), pp. 39–73. New York: Hemisphere.

Hamilton, W. D. (1964). The genetical evolution of social behavior. *Journal of Theoretical Biology*, 7: 1–52.

Hoenigswald, H. M. and Wiener, L. F. (ed.) (1987). *Biological metaphors and cladistic classification*. Philadelphia: University of Pennsylvania Press.

Heyes, C. M. (1995). Imitation and flattery: A reply to Byrne and Tomasello. *Animal Behaviour*, 50: 1421–4.

Heylighen, F. (1999). The necessity of theoretical constructs. *Journal of Memetics– Evolutionary Models of Information Transmission*, 3. [http: //www.cpm.mmu.ac.uk/ jom-emit/]

Hull, D. L. (1988a). *Science as a process*. Chicago: University of Chicago Press.

Hull, D. L. (1988b). Interactors versus vehicles. In *The role of behaviour in evolution* (ed. H. C. Plotkin), pp. 19–50. Cambridge MA: MIT Press.

Hull, D. L. (1995). La filiation en biologie de l'evolution et dans l'histoire des langues. In *Le paradigme de la filiation* (ed. J. Gayon), pp. 99–119. Paris: Editions l' Harmatten.

Hull, D. L., Glenn, S. and Langman, R. (2000). A General Account of Selection: Biology, Immunology and Behaviour. *Behavioural and Brain Sciences*, Cambridge: Cambridge University Press.

Hull, D. L., Tessner, P. and Diamond, A. (1978). Planck's principle. *Science*, 202: 717–23.

Lakatos, I. (1970). Falsification and the methodology of scientific research programmes. In *Criticism and the growth of knowledge* (ed. I. Lakatos and A. Musgrave), pp. 91–196). Cambridge MA: Cambridge University Press.

Lake, M. (1998). Digging for memes: the role of material objects in cultural evolution. In *Cognition and material culture: The archaeology of symbolic storage* (ed. C. Renfrew and C. Scarre), pp. 77–88. University of Cambridge. McDonald Institute Monographs.

Lande, R. (1988). Genetics and demography in biological conservation. *Science*, 241: 1455–1460.

Laurent, J. (1999). A note on the origin of 'memes'/'mnemes'. *Journal of Memetics– Evolutionary Models of Information Transmission*, 3. [http: //www.cpm.mmu.ac.uk/ jom-emit/]

Lumsden, C. J. and Wilson, E. O. (1981). *Genes, mind and culture*. Cambridge MA: Harvard University Press.

Majerus, M. E. N. (1998). *Melanism: Evolution in action*. Oxford: Oxford University Press.

Marsden, P. (1999). A strategy for memetics: memes as strategies. *Journal of Memetics–Evolutionary Models of Information Transmission*, **3**. [http: //www.cpm.mmu.ac.uk/jom-emit/]

Maynard Smith, J. and Szathmáry, E. (1995). *The major transitions in evolution*. New York: Freeman.

Mayr, E. (1982). *The growth of biological thought*. Cambridge MA: Harvard University Press.

Mayr, E. (1983). Comments on David Hull's paper on exemplars and type specimens. *PSA 1982* (ed. P. D. Asquith and T. Nickles), **1**: 504–11. East Lansing MI: Philosophy of Science Association.

Mendel, G. (1869). Versuche über Pflanzen-Hybriden. *Verhandlungen des naturforschenden Vereines in Brünn*, **4**: 3–57.

Pocklington, R. and Best, M. L. (1997). Units of selection in a system of cultural replication. *Journal of Theoretical Biology*, **188**: 79–87.

Portin, P. (1993) The concept of the gene: short history and present state. *The Quarterly Review of Biology*, **68**: 173–223.

Saccheri, I. *et al.* (1998). Inbreeding and extinction in a butterfly metapopulation. *Nature*, **392**: 491–4.

Semon, R. (1904). *Die Mneme als erhaltendes Prinzip in Wechsel des organischen Geschehens*. Leipzig: W. Englemann.

Semon, R. (1914). *The mneme*. London: George Allen & Unwin Ltd.

Speel, H-C. (1999). On memetics and memes as brain-entities. *Journal of Memetics–Evolutionary Models of Information Transmission*, **3**. [http: //www.cpm.mmu.ac.uk/jom-emit/]

Stent, G. (ed.). (1980). *Morality as a biological phenomenon: the presuppositions of sociobiological research*. Berkeley: University of California Press.

Tooby, J. and Cosmides, L. (1990). On the universality of human nature and the uniqueness of the individual: The role of genetics and adaptation. *Journal of Personality*, **58**: 17–67.

Vrba, E. and Gould, S. J. (1986). The hierarchical expansion of sorting and selection: Sorting and selection cannot be equated. *Paleobiology*, **12**: 217–28.

Wanscher, J. H. (1975). The history of Wilhelm Johannsen's genetical terms and concepts from the period 1903–1926. *Centaurus*, **19**: 125–47.

Wilkins, J. S. (1998a). What's in a meme? Reflections from the perspective of the history and philosophy of evolutionary biology. *Journal of Memetics–Evolutionary Models of Information Transmission*, **2**. [http: //www.cpm.mmu.ac.uk/jom-emit/1998/vol2.Wilkinsjs.html]

Wilkins, J. S. (1998b). The evolutionary structure of scientific theories. *Biology and Philosophy*, **13**: 479–504.

Wilkins, J. S. (1999). Memes ain't (just) in the head. *Journal of Memetics–Evolutionary Models of Information Transmission*, **3**. [http: //www.cpm.mmu.ac.uk/jom-emit/]s

Williams, G. C. (1966). *Adaptation and natural selection*. Princeton: Princeton University Press.

CHAPTER 4

Culture and psychological mechanisms

Henry Plotkin

A natural science of culture will take several different forms, any one of which could reasonably claim to be Darwinian. Broadly, they would be of two possible kinds. The first would comprise the claim, and the attendant empirical programme which it mandates, that human culture and the human capacity for entering into culture, is a consequence of evolution. In other words, this approach would be concerned with the evolution of culture, specifically the evolution of the psychological mechanisms that comprise the human capacity for entering into culture. The second deals with how cultures change. Some of these approaches make an explicit commitment to the notion that such transformation in time is the result of the same processes that drive biological evolution (a form of universal Darwinism); others are more concerned with the broader framework of gene–culture coevolution. Thus, this second approach (in its various forms) is focused on cultural evolution, and it too has an empirical programme. Not only are these two broad approaches not exclusive of one another, but a future complete science of culture will require their integration.

Memetics was born out of the second of these approaches and has mostly taken the form of a universal Darwinism. Universal Darwinism—which has its origins in the 1860s with the suggestion by T.H.Huxley that the theory of evolution by selection be extended to explain individual development—has been embraced by a long line of theorists including Darwin himself, William James, James Mark Baldwin, and Karl Popper among others (see Plotkin 1994 for a history). The basic premise of universal Darwinism is that the processes of variation, selection, and the conservation of selected variants, are common to the causes of the transformation in time of a number of complex biological entities and systems. These include not only lineages of species but also changes in parts of the nervous system and immune systems. These common processes, of course, are embodied in entirely different mechanisms in each case. The application of universal Darwinism to culture and cultural change originates with Murdock (1956). Thus, memetics in its contemporary form is part of a long line of thought. However, if memetics is to mature into a successful science then it must equally become part of the enterprise of those whose concern is the evolution of the human attribute of culture, and it needs to be underpinned by a knowledge of psychological mechanisms. Now social scientists seldom deal with mechanism in the way that natural scientists do. For the latter, a mechanism is something that you can touch or taste; it has material substance. For most social scientists, on the other hand, a mechanism— if they think about mechanism at all—is usually a rule that describes an interaction or process. And any natural science approach which has attempted to introduce biology into the study of culture, especially as it is concerned with mechanism, has been consistently criticized, and still is consistently criticized, as being reductionist (usually genetic reductionist) and simple-minded.

My aim here is briefly to refute the charge of reductionism; and then to attempt to save memetics from the second criticism, simple-mindedness, which is justifiably invoked against memetics of a particular kind. This will be done by appealing to psychological mechanisms as the basis for a pluralistic approach to the concept of memes.

Reductionism refuted

Culture is the product of individual human intelligence. Intelligence here is broadly defined not psychometrically but as the capacity of any animal to generate the causes of some of its own behaviour as a result of the activity of dynamic neural network states, thus allowing a degree of behavioural flexibility. Such flexibility, which finds extreme expression in humans, is to be contrasted with the relatively stereotyped responding of unintelligent animals. Their behaviour is proximally caused by immediate stimulation of receptor surfaces, the consequences of which are processed by relatively fixed neural network states with invariant output to effectors, all of which are the products of genes and appropriate developmental conditions.

Intelligence takes many different forms and is widespread in the vertebrate subphylum as well as being present in some other phyla, notably the *Arthropoda*. It is likely that intelligence evolved originally because of the advantages of being able to detect and act on conserved, covarying, events in the world (i.e., causal relations). There is strong evidence (Dickinson and Shanks 1995) that associative learning is one of the mechanisms underpinning human judgements of causality. It is also likely that intelligence first evolved hundreds of millions of years ago, and the original associative learning ability has been elaborated upon to produce the array of learning and cognitive mechanisms that are now the bread and butter areas of psychological study including thought, problem-solving, and decision-making, as well as the more specialized cognitive abilities like language learning.

The evolutionary history of intelligence in its myriad forms is, in detail, unknowable. But what is clear is that the evolution of intelligence marks an important step in the history of animal life. It involved a partial shift of behavioural causation away from genes and development into neural networks. This quite fundamental shift in the causes of some behaviours negates the claim of reductionism in any biological account of the behaviour of intelligent animals (Plotkin 1994). Since culture is a manifestation of the complex and multiple intelligences of humans, no biological evolutionary account of culture can ever be reductionist either in intent or outcome. This applies as much to memetics as to any other school of thought. For the social sciences, reductionism is a fear without foundation and one that no social scientist should entertain. The argument just presented is concerned specifically with behaviours driven by intelligence. However, the argument can and has been widened. Fearful social scientists can be assured that virtually all complex human psychological and behavioural traits are beyond genetic reduction (Sarkar 1998).

Kitcher's rule

As part of a penetrating criticism of human sociobiology, Kitcher (1987) wrote that 'without a serious psychological theory onto which the considerations about cultural transmission can be grafted' no understanding of culture by the natural sciences is possible. Given that culture is, and can only be, a product of human minds, I take this statement to be self-evidently correct, and one which must form a central conceptual plank in the programme to biologize the science of culture. 'Serious psychological theory' presumably refers to the approaches and concepts of contemporary psychological science, and for the contemporary psychological scientist, mechanism is all. A mechanism is an entity with a specific psychological function which has specific characteristics, whose existence is supported by empirical evidence, and which can be, or can potentially be, sited within a particular anatomical structure with characteristics that match the psychological function of the mechanism. For example, a supervisory attentional system, sited in the frontal lobes, whose function is to modulate the activities of a contention-scheduling system, is a specific high level cognitive mechanism whose existence, psychological characteristics, and broad neurological features are supported by laboratory experimentation and neuropsychological case studies (Shallice 1988). The high level nature of the supervisory attentional system means that it certainly is a psychological mechanism that plays an important role in the process of enculturation. It is, however, a mechanism that enters into very many human activities and functions, and may well be present in other species, especially extant great apes.

The example of the supervisory attentional system forces a distinction. On the one hand, there are psychological mechanisms which, in humans, may play a part in the ability to enter into culture, but which are shared with other species which do not have culture. On the other hand, psychological mechanisms also exist that are uniquely human and which there is good reason to believe are essential mechanisms of human culture. This is an important point and needs expanding upon. The basic assumption is that human culture is unique. Other species, notably chimpanzees, display systematic variations across a range of behaviours (Whiten *et al.* 1999) that suggest a capacity tantalizingly close to culture, which one may label as protoculture. However, the characteristics of human culture—such as the sharing by virtually every member of a social group of skills, knowledge and beliefs, and the incessant and cumulative modification of practises and knowledge over many, many generations—is absent in other species (Tomasello *et al.* 1993). Now, because of the complexity of culture, it is almost certainly, and obviously, the case that virtually every fundamental psychological mechanism of humans—including sensation, perception, memory, reasoning, attention, skilled motor performance, motivation and emotion—is involved in that ability of humans to create and enter into culture. Many of these mechanisms, such as memory and attention, are present in other species. But some mechanisms are unique to humans. The distinction which is being drawn is between psychological mechanisms that are common to humans and some other species, and which may well contribute to the distinctiveness of human culture, and mechanisms that are unique to humans and upon which the existence of human culture may hinge. Homing in on the latter is the first step to understanding those mechanisms whose evolution was necessary for the emergence of human culture. The additional step of examining how species-common mechanisms may uniquely contribute to the human capacity for culture is not one that will be followed here.

There are two caveats that should be sounded about this distinction. The first is that chimpanzee protoculture makes the line to be drawn between culture and protoculture a distinctly fuzzy one, as is the line between language and protolanguage. Advances in the study of animal behaviour confound a craving for neat distinctions. Second, it is possible that certain human-specific psychological mechanisms exist which do not play an essential role in culture. However, as a working hypothesis, one can assume that the crucially important mechanisms that we are looking for in fulfilment of Kitcher's rule are those that are either wholly human-specific, or present in other species in barely detectable form.

Definitions of culture

There is an inverse relationship between the importance of definitions and how advanced a science is.When dealing with complex issues over which there is little agreement, definitions really do matter. The social sciences that have pursued the study of culture over this last century were not a unified movement, and were guided by the often competing demands of many different schools of thought. One of the results was literally hundreds of definitions of the phenomenon being studied (Kroeber and Kluckholm 1952; Keesing 1974), and the frequent failure of communication between scholars speaking different languages about different aspects of culture. Such, perhaps excessive, pluralism is partly the result of culture's astonishing complexity, and partly due to method-ological differences in the approach of different schools. Kitcher's rule requires a definition consonant with both the emphasis on psycholog-ical mechanism and the complexity of culture. 'Shotgun' definitions, like Tylor's 'knowledge, belief, art, morals, law, custom, and other capa-bilities and habits acquired by man [*sic*—this is a nineteenth-century definition] as a member of a society including every artefact of cultural behaviour' give no conceptual or methodological toe-holds for anyone wielding Kitcher's rule.

In contrast, Goodenough's 'whatever it is one has to know or believe to operate in a manner acceptable to that society's members' (Goodenough, 1957) is compatible with Kitcher's rule because it provides a focus of sorts on different kinds of psychological mecha-nisms—those concerned with knowledge and beliefs, their sharing, and social acceptance. The thrust of Goodenough's definition is on *shared* knowledge and beliefs, because it is that sharing, actual or simulated, which leads to acceptance and the cohesion that marks a common culture. Shared knowledge and beliefs covers a very wide range of possible means of inducing sharing, as well as a wide range of what it is that is shared. Current psychological theory does not support the notion of some single mechanism underpinning culture as defined infor-mationally and socially by Goodenough.

Different forms of knowledge and belief

There are, of course, many different forms of knowledge and belief, and it is almost ludicrous to have to point this out. Thorndike's (1898) imitation, whereby an act, a literal motor act, is learned through observing the performance of that act by another—'learning from an act witnessed'—is one way of gaining knowledge of one particular kind. Doubts have been expressed about whether imitation is capable of supporting behavioural traditions, at least in non-humans, because of the susceptibility of such behaviour to change by individual learning (Heyes 1993). The difficulties of resolving such seemingly picayunish concerns is the stuff of normal science, of course. But they should not detract from the strong possibility of 'motor traditions', be it in the construction of a stone axe or how most effectively to use a weapon, as being a part of human culture. It is also likely, in my view, that imitation was probably important in the evolution of human culture, and perhaps important specifically in the evolution of language. However, it should be noted that imitation is but one of a cluster of different forms of social learning (Heyes 1994; Heyes and Galef 1996) and as yet there is no evidence as to communality or difference of mechanism between them. It should also be noted that recent reports of protoculture in chimpanzees (Whiten *et al.* 1999) point to imitation as the principal form of information transmission, suggesting that imitation is a form of learning which is not confined to our species. So while I think it can reasonably be argued that imitation has played, and probably still does play, a role in human culture, it is not a central role and it is not exclusively human. We are not going to prise open the secrets of human culture by concentrating our studies on what people do— which is what imitation is about. Imitation, normally and properly defined, is concerned with learning a set of actions. But important as actions are, they are not everything.

Now contrast imitation, a visuomotor form of learning, with the acquisition of language. Language must be distinguished from the imitation of a speech act. Language is the use of a finite number—usually a quite small number—of elements (symbols), to generate a virtually unlimited number of utterances (signals), each of which has meaning. Language is only acquired within an environment of other language users, and there is strong evidence that language is not modality-specific but processed in the same brain regions irrespective of modality of input (e.g., Hickok *et al.* 1998). It is also a form of knowledge that becomes the vehicle for acquiring other forms of knowledge, and at least partly the vehicle for acquiring beliefs. Nothing, absolutely nothing, known at present suggests communality of mechanism between language and imitation. And no one, absolutely no one, disputes the central role of language in human culture.

Then contrast imitation and language with the shared beliefs of social constructions. Following Searle's (1995) analysis of the construction of social reality, I assume that entities which come into being only because of widespread agreement within a culture that these things do exist— things like money, justice, and marriage—are essential features of every culture. The nature of the social constructions, obviously, varies from culture to culture. For example, while the social construction of justice is almost always present in every culture, the agreed basis for justice—whether fairness in the distribution of resources, ownership, kinship, service to the social group, or revenge—does differ between cultures. So vary they will. But what is invariant is the existence of social constructions in every culture. Exactly what psychological mechanisms are necessary for humans to enter into social constructions remains unclear. I have suggested elsewhere (Plotkin 1998) that the 'theory of mind' mechanism—that is, the mechanism that allows the attribution of intentional mental states to others—is essential for participating in social constructions. At present, this stands merely as a hypothesis, but the study of the understanding of social constructions in individuals with impaired theory of mind provides the promise of an empirical test of it.

The main point to be made, though, is this. The imitation of a motor act, the acquisition of a native language, and learning one's culture-specific social constructions have different developmental trajectories. If I am right about the important role of theory of mind to social constructions, then they are also sited in different parts of the brain, and they make different computational demands from one another. Each is based on different psychological mechanisms. It is almost certainly the case that the characteristics each displays in terms of fecundity, longevity, and fidelity of copying are also different in each case, and different precisely because each is based on different mechanisms. The suggestion that 'we stick to defining the [sic] meme as that which is passed on by imitation' (Blackmore 1998), if taken literally, is an impoverishment of memetics for reasons of wanting to maintain copying fidelity. It is an error, I suggest, for at least four reasons.

The first is that if one maintains the accepted definition of imitation, that stemming from Thorndike, then what memetics becomes is a kind of one-dimensional account of culture. It leaves out of the science the complex cognitive mechanisms that are responsible for what the social scientists see as the interesting and complicated features that makes culture such a flexible and complex phenomenon. This really is an error of simplification and the social scientists' nightmare. Limiting memetics this way is to validate the claims of the social scientists that bringing natural science into the study of culture is to oversimplify the issues.

The second error occurs when recognition of the need to retain the feature of cultural complexity leads to the claim that 'all these kinds of learning and teaching (i.e. which are a manifest part of culture or which lead to cultural transmission) require at least the ability to imitate' (Blackmore 1998). It is not clear what this means. But if it refers to a psychological mechanism then the notion of imitation has been expanded beyond the point of meaning. Perhaps this is the price paid for putting forward process-based, substrate-neutral formulations in preference to explanations based on causal mechanisms. I would suggest that 'the ability to imitate' be substituted by 'the presence of a copying process', though that still needs a translation into a specific mechanism.

The third error is the assumption that imitation results in less copying errors and is more rapid than other forms of information transmission. Whether this can be tested is an interesting issue. My guess is that if it can be, it will be shown to be wrong. Teaching someone to serve a ball on a tennis court by demonstrating the correct action is a slow process. Telling them to go and eat at the La Strada on Gourmet's Avenue will result in perfect transmission every time.

The fourth error is the assumption that universal Darwinism always requires high copying fidelity in the same way that biological evolution does. However, other biological systems, like the vertebrate immune system and certain forms of learning, are transformed in time by the same processes of variation, selection and conservation, and propagation of selected variants. Yet, copying fidelity, as with longevity and fecundity, varies across these systems. There is no reason why such variation should not also extend to memetics (Heyes and Plotkin 1989).

If one moves away from the monolithic position of imitation as the basic mechanism of memetics and allows for the existence of different kinds of memes based on different mechanisms, of which imitation can be one, then there is every likelihood that different meme systems will be characterized by differences in copying fidelity, longevity, and fecundity. This would provide a complexity more in line with the complexity of culture, and limit, if not eliminate, the criticism that memetics is simple-minded.

A complex architecture for memes

Mechanism alone is not the only way of differentiating between alternative forms of a meme. Transmission routes (such as between- or intra-generational), and the numbers of sources (or 'parents') that can contribute to the 'memotype' of an individual—factors which have already been investigated by existing models of gene–culture coevolutionary models (Cavalli-Sforza and Feldman 1981; Boyd and Richerson

1985)—will also be reflected in differences in mechanism. So too will another dimension that satisfyingly adds to the complexity of memetics. This is the 'scope' (for want of a better term) of information that is being transmitted, which bears upon rates of transmission and longevity. Consider the simple example of being told that a particular shop is selling computers at bargain prices, and then passing that information on to others. Let us pass over the considerable problems of how the brain and psychological states of the first informant can be considered to have been replicated in the receiver's brain, and simply accept that all those told, head for the same-named shop with the same expectations of computer bargains. Hence, operationally at least, the information inside each person's head is sufficiently similar to guide identical expectations and behaviour—and hence might be considered to have been replicated in this broad sense. In fact, what is being replicated in this case is 'simple' and can be stored, transmitted, and replicated as a name and a location. The same applies to the information that a specific restaurant is worth a visit or a dentist is to be avoided. Such information is the 'small change' of culture, based on the episodic memories of individuals. There is no reason not to consider them memes, but they have the characteristic of being informationally narrow in scope. That is, they are very specific—this shop, that restaurant. Their shelf-life is also relatively short. Tomorrow the bargains will be elsewhere and a new and better restaurant will have opened. We are constantly exposed to such situation-specific memes which form a kind of froth to daily social life. These are, to maintain the metaphor, surface memes.

Surface memes, however, are dependent on higher order memories and knowledge structures—referred to in the psychological theory of an earlier age as schemas (Bartlett 1932), and more recently as frames (Minsky 1975), scripts (Shank and Abelson 1977), and memory organization packets and thematic organization points (Shank 1982). These too are memes, but memes of much wider scope informationally, and of much greater longevity, with transmission normally restricted to just once in a lifetime. For example, the higher order knowledge structure associated with the notion of shops is an aggregate, complex, and abstracted characterization of places that one goes to where an array of goods are on display which can become one's own property in exchange for money. Of course, the characterization is usually more complex and will take in increasingly surface features such as the knowledge that some shops specialize and others do not, and that credit cards and cheques can substitute for cash. These higher order knowledge structures are also closely interwoven with others, like money.

Such higher order knowledge structures are acquired by every child in every culture through the long process of enculturation by which we are all inducted into the knowledge of how our culture works and what its beliefs and values are. The information acquired is of wide scope informationally, yet circumscribed. Shops are different things from schools, and both are different from prisons. They are also culture-specific, because many cultures have none of these things, while our culture has no place for higher order knowledge structures relating to, say, the behaviour of animals and the bearing that animals have on our welfare. The transmission of higher order knowledge structures is smeared out over a considerable period of time, yet the replication achieved is probably just as accurate as is an imitated motor act. We all share the same higher order knowledge structures concerning what shops or schools are. Transmission occurs at the same rate as genetic transmission—i.e., once in a lifetime. These deep-level, culture-specific memes are essential for the existence of surface memes. They are not acquired by imitation but by a complex process of construction and integration. Native language learning and the acquisition of the social constructions characteristic of a culture share some of the characteristics of deep-level memes in terms of rates of transmission and longevity of memes, but the mechanism of transmission and replication might be quite different.

It is, of course, important to maintain the distinction between mechanism and the product of a mechanism. Surface- and deep-level memes can be identified with, and are the products of, specific psychological mechanisms. Those mechanisms are themselves products of another set of mechanisms that we collectively refer to as evolution, and hence are universal to all humans. If memetics is to become a mature science based on the understanding of mechanisms as causal explanations, then at least some of its practitioners need to become involved in what is *the* fundamental issue in cognitive psychology at present, and which is likely to remain so for some time. This is the extent to which human cognition is based upon domain-specific cognitive modules which have evolved as predispositions to acquire specific kinds of information, and the general process approach which is antithetical to the modularity thesis and more akin to a *tabula rasa* view of the human mind. The resolution of so theoretically deep an issue in psychology is bound to have repercussions on memetics.

One of these repercussions is that if the modularity position prevails, then all humans, irrespective of culture, are predisposed to acquire memes that cluster about those predispositions, whose origins lie in those selection pressures that were constant in human evolution. That life was consistently lived within small social groups was one of the very few constants the existence of which we can be reasonably confident about. Thus, the psychological predispositions that are the very basis of meme production may be tuned to specific features of the social world—like the control of social interactions, the division of resources within the group, group defence, relationships between the sexes, adult–child relations, responses to outsiders, and shared large scale causal attributions (ontology, metaphysics). These are what might be the foci of deep-level meme clusters. Although this is speculative, it is also empirically answerable—one of the activities that David Hull (this volume) refers to as 'doing memetics'.

Conclusion

Acceptance by social scientists is not the acid-test of attempts to naturalize the science of culture. However, social scientists do know more about culture than biologists because they have been studying culture for about as long as biologists have been studying evolution. One of their messages is that cultures are complex entities based not on tying shoe-laces or using forks but on knowledge, beliefs and values like ritual observances, origin myths, the pursuit of happiness, obedience to God's law, and money markets (see Bloch, this volume). The notions of universal Darwinism, replicators and interactors as the basic concepts of memetics may well prove a fruitful approach to the understanding of culture; even more importantly, it might provide one of the conceptual bridges between the biological and social sciences that we are all looking for (Plotkin 2000). But imitation is not a process, it is an ill-understood mechanism (see Laland and Odling-Smee, this volume). And basing a science of memetics on the single mechanism of imitation—which, it should be noted, is a form of general process approach to cultural cognition—will not deliver as an explanatory basis for culural complexity, and will lay itself open to ridicule by social scientists. Nowhere is Occam's Razor more misplaced than in a science of culture.

References

Bartlett, F. C. (1932). *Remembering*. Cambridge: Cambridge University Press.

Blackmore, S. (1998). 'Imitation and the definition of a meme.' *Journal of Memetics– Evolutionary models of Information Transmission*, 2. [http://www.cpm.mmu.ac.uk/ jom-emit/1998/vol2/blackmore_s.html]

Boyd, R. and Richerson, P. J. (1985). *Culture and the evolutionary process*. Chicago: University of Chicago Press.

Cavalli-Sforza, L. L. and Feldman, M. W. (1981). *Cultural transmission and evolution: A quantitative approach*. Princeton: Princeton University Press.

Dickinson, A. and Shanks, D. (1995). 'Instrumental action and causal representation.' In *Causal Cognition*, (ed. D. Sperber, D. Premack and A. J. Premack), pp. 5–25, Oxford: Clarendon Press.

Goodenough, W. H. (1957). 'Cultural anthropology and linguistics.' In *Report of the 7th Annual Roundtable on Linguistics and Language Study* (ed. P. L. Garim), pp. 167–73, Washington DC: Georgetown University Press.

Heyes, C. M. (1993). Imitation, culture and cognition. *Animal Behaviour*, **46**: 999–1010.

Heyes, C. M. (1994). Social learning in animals: categories and mechanisms. *Biological Reviews*, **69**: 207–31.

Heyes, C. M. and Galef, B. G. (1996). *Social learning in animals: The roots of culture*. London: Academic Press.

Heyes, C. M. and Plotkin, H. C. (1989). Replicators and interactors in cultural evolution. In *What the philosophy of biology is.* (ed. M.Ruse), pp. 139–62, Dordrecht: Reidel.

Hickok, G., Bellugi, U. and Klima, E. S. (1998). The neural organization of language: Evidence from sign language aphasia. *Trends in Cognitive Science*, **2**: 129–36.

Keesing, R. M. (1974). Theories of culture. *Annual Review of Anthropology*, **3**: 73–97.

Kitcher, P. (1987). Confessions of a curmudgeon. *behavioural and brain sciences*, **10**: 89–97.

Kroeber, A.L. and Kluckholm, C (1952). Culture: A critical review of the concepts and definitions. *Papers of the Peabody Museum of American Archaeology and Ethnology*, **47**: 1–22.

Minsky, M. L. (1975). A framework for representing knowledge. In *The psychology of computer vision* (ed. P. W. Winston), pp. 211–77, New York: McGraw-Hill.

Murdock, G. P. (1956) How culture changes. In *Man, Culture and Society* (ed. H. L. Shapiro), pp. 247–60. Oxford: Oxford University Press.

Plotkin, H. (1994). *Darwin machines and the nature of knowledge*. Cambridge MA: Harvard University Press.

Plotkin, H. (1998). *Evolution in mind*. Cambridge MA: Harvard University Press.

Plotkin, H. (2000 in press). Evolution and the human mind: How far can we go? In *Naturalism, Evolution and Mind* (ed. D. Walsh).

Sarkar, S. (1998). *Genetics and reductionism*. Cambridge, Cambridge University Press.

Searle, J. R. (1995). *The construction of social reality*. London: Allen Lane.

Shallice, T. (1988). *From neuropsychology to mental structure*. Cambridge: Cambridge University Press.

Shank, R. C. (1982). *Dynamic memory*. New York: Cambridge University Press.

Shank, R. C. and Abelson, R. (1977) *Scripts, plans, goals and understanding*. Hillsdale NJ: Erlbaum.

Thorndike, E. L. (1898). Animal Intelligence: an experimental study of the associative process in animals. *Psychological Review Monographs*, 2: 1–109.

Tomasello, M., Kruger, A. C. and Ratner, H. H. (1993). 'Cultural Learning.' *The Behavioural and Brain Sciences*, 16: 495–511.

Whiten, A. *et al.* (1999). Culture in chimpanzees. *Nature*, 399: 682–685.

CHAPTER 5

Memes through (social) minds

Rosaria Conte

A social cognitive perspective on memetics

In this chapter, I take a social cognitive perspective on memetics. By this I mean the study of the cognitive requirements for intelligent but limited autonomous agents to engage in social (inter)action (Conte 1999). More specifically, by a cognitive process, I mean a process that involves symbolic mental representations (such as goals and beliefs), and which is accomplished by means of the operations that agents perform on these representations (reasoning, decision-making, etc.). A social cognitive process is a process that involves social beliefs and goals, and which is effectuated by means of the operations that agents accomplish on social beliefs and goals (like social reasoning). Finally, a belief or a goal is social when it mentions another agent and possibly one or more of her mental states. (For a discussion of these notions, see Conte and Castelfranchi 1995; Conte 1999).

This type of social cognitive approach is receiving growing attention within several subfields of the so-called *sciences of the artificial* (Simon 1956)—in particular intelligent software agents, Multi-Agent Systems, and Artificial Societies. Unlike the 'theory of mind' (cf. Leslie 1992), this approach is aimed at modelling and possibly implementing systems which act in a social (whether natural or artificial) environment. Whilst the theory of mind focuses on one important aspect of social agency— social beliefs (what agents know about others)—the approach presented here is aimed at modelling the various mental states (including social goals, motivations, obligations) and operations, such as (social) reasoning and decision-making, necessary for an intelligent social system to act in some domain[1] and to influence other agents (through social learning, influence, and control).

It has been formally shown (Conte and Castelfranchi 1995; Conte *et al.* 1998; Castelfranchi *et al.* 1999) that a social cognitive approach is needed to account for the mental implementation of social institutions (the so-called Micro–Macro link). Social cognitive processes are essential to explain how social or legal norms are observed or violated, how social control is produced, and so on. Social reinforcement (Bandura 1971), by means of which actions corresponding to the norms are reinforced while actions diverging from the norms are punished, is an insufficient mechanism. First, it does not account for norm recognition. In order to tell that something is a norm, agents need a mental representation of it, since actions may have costs independent of punishment, and achieve goals independent of reinforcement. Action costs do not always (and are not always expected to) discourage agents from the corresponding actions. Only a subset of them, namely those which derive from norm violation, do so. For example, the cost of legal parking is sometimes not much lower than a fine for illegal parking, and yet illegal parking is certainly discouraged, while costly but legal parking is not. How to tell the difference without a representation of sanction as a special cost of action, deriving from norm violation?

A second difficulty resides in norm conflicts. Complex societies entail a growing number of interfering institutions, with their corresponding norms and precepts. Agents can identify such conflicts and solve them in a (globally) useful way only if they are able to reason upon the norms. Otherwise they will simply choose the most reinforcing action. Finally, how to explain social control without a representation of the norm? How could agents reinforce one another to obey the norms, if they had no ideal representation to which to compare others' behaviours? Moreover, why should they want to do so, if they had not formed a

---

[1] Furthermore, if the theory of mind is focused on natural systems, the present approach is often involved in the implementation of artificial agents. This may appear as an advantage of the theory of mind over the social cognitive approach to agent systems. But the computer implementation of an agent model offers a testbed for assessing whether that model is internally valid and whether it is sufficiently complete to account for the realization of the target phenomenon. A model's internal validity (unhappily called 'verification' by computer scientists) can be ascertained also by non-computational means (think of mathematical methodologies). But a model can be internally valid and at the same time severely incomplete, or mutilated, and therefore it will not suffice to realize the phenomenon under study. The 'theory of mind' addresses the question as to how social agents form social beliefs but does not investigate how they achieve their goals about and through others. The present approach is aimed to model this fundamental aspect of social agency.

normative will of some sort? Social control is crucial in the transmission of social norms, conventions, rules, and customs. Thus, I claim that social cognition is fundamental to an understanding of the transmission of norms and other institutions.

Norms and other social institutions are systems of beliefs, prescriptions, and rules—or complex memes—which emerge and spread thanks to social and cognitive processes, and which interact with other components of the culture. A social cognitive model helps to explain the emergence and evolution of social institutions as well as other aspects of culture. The evolutionary algorithm (Dennett 1995) operates on culture through the mental processes and capacities of social agents. A fundamental property of social cognitive agents is their (limited) autonomy. In given societies (in particular, human and simulated information societies), agents are autonomous: they decide whether to accept or reject external requests and inputs. They decide whether to observe or violate norms, whether to retain or discard existing cultural inputs. Thanks to social cognitive processes, agents may recombine existing, possibly inconsistent inputs or contradictory requests (e.g., conflicting norms) and therefore contribute to their evolution. Thanks to the same processes, agents operate on any other aspect of culture: they select, recombine and contribute to the evolution of (systems of) beliefs, customs, habits, and rules of practice.

Therefore, memetics can account for culture if it explains

- How memes operate through and across the minds of the agents, and how minds operate on memes.
- What a memetic mind is, or what are the requirements of a memetic mind. In the present view, a memetic mind is a *social* one. Later on in the chapter, I will clarify what a social mind is.

In the rest of this chapter, I will defend this fundamental claim by referring to social cognitive models, on one hand, and computational and simulation-based evidence, on the other. In the next section, some important advantages of memetics, as they are perceived by a non-expert in the field, are recalled, and some missed or weak points are addressed. These essentially amount to an inadequate or insufficient explanation of why and how memes replicate. Existing theories or speculations will be found insufficient or useless, essentially because they have no conceptual and theoretical instruments to deal with beliefs and their transmission. In the following section, contributions which could be offered by Multi-Agents Systems and Agent-Based Social Simulation to the development of memetics are examined. Next, a model of a social cognitive agent will be briefly outlined. In the successive sections, the model is shown to address some fundamental objectives of a memetic theory: to explain *how* memes are transmitted; to formulate (working) hypotheses and predictions about *to what extent* given memes will replicate; to formulate hypotheses about *which* memes will be more likely to replicate given competition or interference among distinct memetic processes; and to investigate and foresee which *effects* memetic transmission will have on social and collective behaviour. Finally, some fundamental memetic notions will be redefined in terms of this social cognitive model. A recapitulation and some final remarks conclude the chapter.

Memetics: Hits and misses

There are several good points about memetics, which it is useful to recall. First, the memetic approach is a *foundational* one: its main purpose is to understand the elementary principles of cultural transmission.

Second, it shares the advantages of any *evolutionary* approach: it is heuristic in nature, not only stimulating novel interpretations or reconstructions (see Hull, this volume) of cultural phenomena, but also allowing for new research questions to be addressed (such as differences and similarities between different processes of cultural transmission), or for old questions to be reproposed. (For example, What are the mechanisms for the propagation of memes? What are the roles of imitation, social learning, or social facilitation in memetic processes?) At the same time, it allows these new interpretations to be anchored on the firm ground of selection mechanisms. Finally, it allows us to bridge the gap between phenomena and entities (culture, mind, and organisms) which may appear to be incompatible.

Besides, memetics is intrinsically *interdisciplinary*, bringing together biologists, philosophers, anthropologists, and evolutionary psychologists (although cognitive scientists are presently under-represented in this new field).

Another important point about memetics is that it lends itself to *computational* and *simulation-based* modelling of cultural phenomena. The simulation-based study of social phenomena has already proved fruitful in promoting both the methodological development of the social sciences (cf. Gilbert and Doran 1994; Gilbert and Conte 1995; Conte *et al.* 1997; Gilbert and Troitzsch 1999), and a cross-fertilization between agent theory and social theory (see Sichman *et al.* 1998). Analogously, memetics will have much to gain from a closer interplay with computational modelling.

Finally, memetics is dealing with a wide range of interesting issues, from the survival of institutional concepts (de Jong 1999), to the evolution of financial markets (Frank 1999), and the propagation of social pathologies (Preti and Miotto 1997). The emphasis laid on these phenomena within memetics is certainly valuable. Other social issues of equivalent importance—such as the emergence and spread of social conventions and norms, which are still poorly understood—might equally profit from a development of this field.

In sum, memetics appears to represent a fundamental scientific opportunity for a study of cultural and behavioural transmission.

However, this field is also deficient with respect to the treatment and definition of *memetic agents*. In memetics, agents are essentially viewed as 'vectors' of, rather than actors behind, cultural transmission. This inadequate understanding of the role of agents has a number of disadvantages from the memetic point of view as well—that is, in view of an adequate understanding of the memetic process. Let us see why.

The view of agents as vectors of cultural transmission arises from an *insufficient understanding of the autonomy of (memetic) agents*. The autonomy property has important implications: autonomous agents play a crucial role in the cultural application of the evolutionary algorithm. Of course, memeticists acknowledge that agents may misperceive and re-elaborate memes. But this view is still insufficient. It does not (at least explicitly) account for the decision-making implied by the process leading from perception to belief formation. Between an autonomous agent receiving a given input and its forming a belief (possibly corresponding to the input), a fundamental process takes place—namely a decision-making process—which includes several steps (see p. 91–2). This process is therefore relevant in determining which (set of) inputs will be retained and which will be discarded (see the notion of agents' 'acceptance' of cultural units, as used in Cavalli-Sforza and Feldman 1981).

A consequence of this *inadequate view of the memetic agent is the typical memetic account of the mechanisms of memetic transmission.*

Essentially, memetic transmission is explained in terms of imitation (Dawkins 1976; Blackmore 1999). However, this is only one mechanism among others, which include social learning, goal adoption, social and norm-based influence and control, or conformity. The characteristics of these mechanisms affect the features of the memetic processes and can be used to make hypotheses about transmissibility.

Two major problems arise from the existing views about memes' reproductive success. First of all, these views are 'substantialist': memes are said to replicate because of their characteristics. Dawkins (1976), for example, proposes that the 'psychological appeal' of certain beliefs explains their reproductive success (like the religious ideas about hell). But the notion of psychological appeal is equivalent to the probability that a meme will be accepted, and as such adds nothing to the account. Of course, beliefs that are more likely to be accepted will survive and reproduce more than those that are not likely to be accepted! The question, of course, is what makes a belief more or less acceptable than another? Again, a theory of the social agent and of the criteria on which this agent selects among candidate beliefs is essential.

A complementary hypothesis suggests that memes are useful because they spread. According to this hypothesis, the memes' success depends upon the mechanisms and processes of transmission, rather than upon their contents. To investigate specific mechanisms of transmission would therefore allow us to highlight the reasons for the reproductive success of memes.

A final problem concerns the *mental implementation of memes.* In the memetic literature, the mental implementation of memes is often equated to storing or memorization (cf. Rhodes, 1999), and sometimes to the intuitive but yet vague notion of mental 'harbouring' (Dennett 1995). What does this mean? How are things represented in the mind, or, better, how are beliefs held? This is a challenging question that requires a cognitive answer. The crucial issue is not whether memes reside in the brain or not, for it is undeniable that memes also reside outside the brain. Not only artefacts and products, but also behaviours, represent memes. The real problem is how a meme is implemented in the mind (as a belief, a goal, or an obligation; and if it is a belief, whether it is a social belief, and for which reason it was formed, to what extent it is believed, and how it is believed), since this also tells us a lot about how that meme will travel in the social space.

Multi-Agent Systems and Agent-Based Social Simulation

Within the Sciences of the Artificial, there are several fields which have a lot to say to memetics: the field of software agents, and in particular Multi-Agent Systems (MAS), and the field of Agent-Based Social Simulation (ABSS) or Artificial Societies.

Contributions from MAS are both theoretical and methodological. As for theoretical contributions, in the last decade or so, MAS scientists have been working out models of autonomous intelligent agents (Wooldridge 1999) as software systems endowed at minimum with:

- Proactive-ness, or the capacity to pursue goals.
- Autonomy, or the property to act independent of the programmer's or user's direct intervention.
- Sociability, or the competences necessary for interacting with other agents, whether software or user agents.

In a stronger sense, the intelligent agents implemented in MAS are also cognitive agents, endowed with mental states and the capacity to manipulate them. A classical example of this types of agents is the so-called BDI architecture (e.g., the one first proposed by Rao and Georgeff 1991). A BDI agent is characterized by mental states for Beliefs, Desires, and Intentions, and is able to reason, plan, and take decisions on them.

MAS formalizes and implements agents that can cooperate or coordinate themselves in joint activities in several domains of application (such as air traffic control, military defence, robotics, personal assistance, education, or entertainment), or in negotiations (such as economic transactions in electronic marketplaces). In MAS, social agents are increasingly viewed as complex systems in which several types of interrelated mental states (goals, beliefs, obligations, intentions, commitment, etc.) are formed and account for many social activities. Even in the field of electronic commerce (Sierra forthcoming), recent developments show that software agents used in economic transactions must be guided by morals and conventions, and must have representations of electronic institutions in order to be really helpful, trustworthy, and accepted by the user. On the one hand, current MAS models are particularly concerned with increasing the flexibility and adaptativeness of software agents (cf. Weiss 1999). On the other, flexible software agents need a variable degree of autonomy (adjustable autonomy). At the same time, they must be able to adapt to unpredictable changes in the environment, and therefore modify their plans, generate new mental states, and learn from and monitor others with whom they are cooperating. MAS can be of great help in providing models of the agent properties which are needed for flexible social interaction.

Furthermore, on the methodological side, MAS can provide agent and multi-agent platforms for modelling and observing social phenomena. A BDI platform upon which cognitive agents can be implemented (DESIRE) is now being used for simulating the spread of negotiation conventions (cf. Castelfranchi *et al.* 1999).

With regard to social simulation, this has a longer tradition in the computational study of phenomena of social propagation (cf. Gilbert and Troitzsch 1999), and in particular of the spread of opinions and conventions (cf. SITSIM; Rockloff and Latané 1996). Presently, a cross-fertilization between MAS and social simulation (ABSS; Agent-Based Social Simulation) is taking place.[2] Traditional social simulation was based on very simple, weakly autonomous agents (such as Cellular Automata). Recent developments have imported more intelligent agents from the AI domain (cf. Doran 1994), from MAS (Sichman *et al.* 1994) and evolutionary and learning agents from genetic algorithms, neural nets, and Artificial Life (for one example, see Cecconi and Parisi 1998). This hybridation provides new opportunities for memetics: memetic phenomena can be observed in artificial societies with learning and evolutionary agents, as well as with intelligent agents. One auspicious development is that learning and intelligent agents will merge to a greater extent than has been the case so far.

A model of limited autonomous agents

What kind of agent is a social intelligent agent? Essentially a limited autonomous agent. But what is meant by this? Let us start by defining autonomy, and then we will proceed to characterize limited autonomy.

In a rather generic sense, an autonomous agent is a self-interested agent. In a more specific sense, an autonomous agent is one which has internal criteria to select among inputs. Inputs might generate two types

[2] See the homepage of the anonymous Special Interest Group within the European Network of Excellence 'AgentLink': http://www.cpm.mmu.ac.uk/~scott/abss/ABSS-SIG.html.

of candidate mental representations: beliefs and goals. An autonomous agent is therefore characterized by a 'double filter architecture', allowing both beliefs and goals to be selected (cf. Castelfranchi 1997). These two filters are sequential, but at the same time they allow for an integrated processing of mental representations.

Filtering beliefs

Thanks to this filter, agents have control over the beliefs they form. This filter is rather complex and implies that a number of tests be executed over a candidate belief against several distinct criteria. These are pragmatic or epistemic criteria.

Epistemic criteria include

- *Credibility*, with agents controlling, among other properties, the *coherence* of candidate beliefs with previous beliefs, the *reliability of the source* of the candidate belief: agents accept information from other agents (Gricean principles) provided they have no reasons to doubt their sincerity or competence.

- An interesting epistemic criterion is *non-negotiability*, or the Pascal law. To believe or not is a 'decision'. However, it cannot be made in view of one's pragmatic utility, but only in view of one's epistemic utility. In social interaction, we cannot use threats (*Argumentum ad baculum*) or promise to make people believe something. The difference between persuading to do and persuading to believe is crucial. Since beliefs control goals, this represents a further protection of agents' autonomy.

Pragmatic criteria concern *the reasons for believing something*. Typologies of beliefs generally concern the format of their representation (declarative, procedural); the degrees of certainty; the levels of nesting (one can believe something, without believing that one believes . . .). These typologies are now very well known. Perhaps it is less obvious that beliefs may have a different 'status' in the mind according to the motives for acceptance. The language provides a rich vocabulary: superstitious belief, creed, faith, doctrine, postulate, axiom, principle, conception, idea, view, opinion, and many others. These beliefs vary on several, often quantitative, dimensions, such as certitude (subjective truth value), retractability (how likely a belief will be modified), connectivity (how much a belief is connected with other beliefs), and so on. One interesting quantitative dimension is the 'force' of beliefs (cf. the role of this dimension in the Social Impact Theory; cf. Latané 1981): beliefs vary on how strongly they are held. This is related to certitude, but also to the motives of acceptance, which may lead the agent to ignore the belief's truth value. The kinds of pragmatic criteria, then are many. For example:

- *Self-protection* and *self-enhancement*: agents may be led to accept one among several competing beliefs because of the belief's positive effect on their self-esteem or self-concept.

- *Commitment* to a given (set of) belief(s) provides one important reason for accepting further, consistent beliefs, in spite of or independent of incompatible evidence: agents that accept beliefs out of commitment do not check their truth value.

- *Hypothetical and counterfactual reasoning*: beliefs may be (transitorily) held as means to reason and carry out operations (demonstrations, proofs, experiments). A good example is the priest accepting the atheist's point of view in order to dismantle it.

- *Communication*: A psychotherapist may 'accept' the delusions of her patient in order to communicate with him and give a clinical sense to his fantasies; here the goal is not to carry out a counterfactual argument, but to understand the meaning of the delusions.

- *Empathy*: agents may want to share the views of their close connections.

- *Risk-taking* or *gambling*: agents may participate in lotteries, accepting one alternative and investing (money) in it. In such a case, agents will hold an uncertain belief but behave as if it were certain.

- *Prudence*: agents may accept uncertain information (e.g., rumours, gossip, even calumnies) and behave as if they were certain. Unlike the preceding situation, a risk-minimization strategy applies.

Filtering goals

There are at least two fundamental tests which are performed on goals (see Figure 5.1):

1. *Self-interested goal-generation*. An agent is autonomous if and only if whatever new goal it comes to have, there is at least another goal of that agent, for which, in the agent's beliefs, the former is a means.

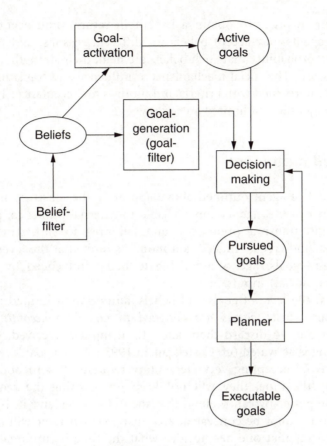

Figure 5.1. The 'double filter' architecture.

2. *Belief-driven goal-processing.* Any modification of an autonomous
   agent's goals can only be allowed by a modification of its beliefs.

Both of these filters have interesting social and memetic consequences.
First, agents' minds are modified thanks to a process of belief-formation
or belief-revision. Second, belief-formation and belief-revision are
decision-based, selective processes. With Cavalli-Sforza and Feldman
(1981), we will speak of beliefs' 'acceptance'. A memetic process is a
process interspersed with decisions taken by the agents involved. But
a decision-based process is not necessarily explicit and reflected on:
mental filters do not necessarily operate consciously, so agents may not
be able to report on them. Third, agents will never accept beliefs under

threat, or in order to obtain a benefit in return (non-negotiability). Fourth, agents may accept beliefs for different reasons, and these will affect the probability that such beliefs are held, their strength, and their transmission. The social mechanisms of influence and transmission are strongly intertwined with criteria and motives for acceptance. This leads us to the question of limited autonomy.

Limited autonomy

The model of agent outlined above appears rather abstract and unrealistic. In real life, agents appear liable to external influence, prone to accept and transmit prejudices, and fall prey to superstition, false doctrines, and creeds. Indeed, autonomy is limited at the level of both beliefs and goals: agents are liable to being influenced by external (including social) inputs.

Both at the level of goals and beliefs, autonomy is limited in a very elementary sense: *agents are designed to take into account external inputs*, if only to discard them later. If an input is received, the filter processing is activated (cf. Castelfranchi 1997). At the goal level, agents cannot avoid accepting very elementary requests. If somebody asks a passer-by the time, they will not keep on ignoring the request. At most, the passer-by can pretend that she did not perceive it. But if this perception cannot be concealed, any answer whatsoever will be given, if only to say that one has no idea what the time is (minimal level of adoption). Of course, this type of influence is rather superficial and ephemeral. But it paves the way for other more relevant types of influence. Obviously, agents' autonomy is limited because they are not always self-sufficient. They may need the help of other agents to achieve their goals (social dependence), and this causes agents to adopt others' goals and to accept their requests. However, one's adoption of others' goals will always be a means for the achievement of one's own goals (e.g., through social exchange or cooperation). In turn, these social actions favour the transmission of beliefs, including action plans, techniques, procedures, rules, conventions, and social beliefs. Finally, agents' autonomy is limited by norms, which are aimed at regulating agents' behaviours. But agents may accept or reject norms, comply with or violate them, always according to their internal criteria for acceptance.

At the level of beliefs, agents' liability varies depending upon the type of beliefs. For example, social agents are strongly permeable to social evaluations, rumours, gossip, even calumnies (Benvenuto 2000). Rumours and gossip are accepted for prudence, and, as we will see, this favours their spread. Indeed, these are important phenomena of memetic transmission—which spread social labels, stigmas, and prejudices, but also reputation, social hierarchies, and other institutions.

Requirements of memetic agents

Two fundamental properties or capacities are usually indicated as essential for cultural transmission: communication (Donald 1991; Gabora 1997), and/or imitation. But neither is a necessary nor sufficient property for memetic processes to occur.

Communication

Memes are not necessarily transmitted through communication. Often, social influence is passive. But even when it is active, it is *not* necessarily communicated. If I want others to believe that I will be staying at home (in order to discourage burglars) while in fact I am about to go out, I can leave the light on. This action is social (directed to modify others' mental states), but it is not communicated (for, otherwise, it would not be efficacious!). A nice example of a non-communicative memetic transmission is the use of 'empty boxes': children are often entrusted with (what they do not know to be) empty boxes (or opaque words). A child must transmit an 'empty box' to an adult in order for the latter to understand that the task was assigned only to send away the child.

Imitation

Imitation is seen as an essential component of memetic processes. Indeed, Dawkins (1976), first, and Blackmore (1999) later, defined memes as units of imitation. Blackmore makes an interesting point when she distinguishes a memetic process (replication) from a non-memetic one (reproduction). However, such a difference is yet unclear because the notion of imitation is not thoroughly convincing. Although extremely important, imitation, is actually one of the 'bad words' of the behavioural sciences. No satisfactory model of imitation has been worked out so far, although developmental psychologists and ethologists have long been trying to define and operationalize it.

The notion of imitation proposed by Thorndike (copying another's behaviour by observing it) to which Blackmore (this volume) refers, is neither necessary nor sufficient for a memetic process to occur. It is unnecessary, because it is a behavioural notion: people observe behaviours, or products, but 'copy' rules, beliefs, intentions, tastes, and standards. But Thorndike's notion is also insufficient, as shown by several forms of automatic contagion (seen above; but see also Marsden 1998), where behaviour is copied automatically without memes being transmitted. Blackmore actually endeavours to make a step forward. She suggests that imitation, and consequently memetic transmission, occurs when a *novel* behaviour is copied. But again, counter-examples abound: people often acquire new behaviours *automatically*, as happens when one can not help using a foreign accent or one's neighbour's tics. As Laland and Odling-Smee (this volume) observe, to imitate often implies learning a new context (and I would add, a new use or meaning) for an old behavioural pattern. The tricky question then is how agents infer these things from observing behaviours. On the other hand, some examples of memetic transmission do not seem to require or rely on imitation. The Simmel effect occurs thanks to a complementarity between the attitude to conform to a minority and the attitude to differentiate from the majority: both these memes spread in a circular way, each reinforcing the other.

This is not to deny the importance of imitation, but to say that *the memetic individual is more than an imitator: imitation implies sociality, but not the other way round!* A theory of imitation requires that mental mechanisms be investigated. In particular:

- *Whom*: what is the target of imitation?
- *What*: which aspects of others' behaviours are imitated?
- *How*: how to infer mental states (memes) from behaviours?

Social competence

If memetic agents do not rely only on communicative and imitational capacity, what else do they need? Here, it is hypothesized that *memetic processes more generally require the evolution of several social cognitive capacities*. As I said in the introduction, a social cognitive capacity includes the capacity to form social beliefs and goals, as well as the capacity to reason about and decide upon them.

Memetic agents must be endowed with the capacity to accept inputs from others, form candidate mental representations, process them, and decide whether to accept, discard or modify them. On one hand, this capacity determines whether and to what extent the propagation of a given social phenomenon will also cause the propagation of memes in the social space. On the other hand, it accounts for specific features of the transmission process, and in particular its stability. As Dennett (1995) puts it, the future of memetics as a science does not depend on the likelihood that memes will be identified in the brain, but rather on the extent to which the reasons and processes which lead to memes being 'harboured' (read, implemented) in their minds are uncovered. And this is precisely where cognitive science can usefully contribute to memetics.

Behaviour and mental representations: The utility of the approach presented in this chapter

So far, I have argued that a memetic transmission of behaviours requires the formation of beliefs and goals in the recipient of transmission. But, obviously, these beliefs and goals may not coincide with those of the actor of transmission. We will say that mental states leading to the same behaviour are *equifunctional*. A crucial aspect of memetic transmission lies in the role played by equifunctional mental states. One could argue that mental states are irrelevant in a model of behavioural transmission: if the same behaviour spreads through a population, the underlying mental states must have been at least equifunctional, if not identical, and that is all one needs to say about them. But this argument is essentially wrong and bears negative consequences for an adequate theory of cultural and behavioural transmission.

In this section, I will endeavour to show the utility of the approach presented so far for memetics. In particular, I will argue that a number of questions of memetic theory cannot be solved without analysing the underlying social cognitive processes among limited autonomous agents. These are

- *how* do memes spread,
- *to what extent* they spread (working hypotheses about transmissibility of memes),
- *which memes* spread, given an interference among distinct memetic processes,
- *which consequences* can be expected from a given memetic process.

How memes spread

In the memetic literature, memes are said to spread essentially by means of imitation. However, this is only one among many possible social mechanisms which are responsible for memetic transmission. First, on the recipient's side, several types of mechanisms might apply. For example, we monitor others (Sherif 1936) in order to check how they perceive a given situation. But this may be based on pre-existing representations, for example, norms: we check others' behaviours to know which norms they are applying (Conte and Dignum forthcoming). Conformity is a form of social monitoring based upon the goal to be like (given) others. In social learning (in the sense defined by Bandura 1971), we learn proper or moral behaviour through social reinforcement. In social facilitation, especially in the ethological sense (Laland and Odling-Smee, this volume), one agent observing another may 'discover' a new routine, procedure, or a new effect of a known action. But memes also spread thanks to active social influence, such as manipulation (hidden influence) and persuasion, or direct and explicit communication.

Each of these mechanisms can have different memetic effects. For example, norm-based monitoring may be expected to have deeper and more stable impact than conformity. While the latter is relativized to others' behaviours, the former is based on observing others' behaviours, but is not relativized to it: once a given norm has been identified through others' behaviours, agents will use the norm itself rather than others as a criterion for controlling their own behaviours. On the other hand, conformity may have a stronger impact on one's behaviour than social reinforcement, because the former is based on one's will to modify one's behaviour, while the latter is thoroughly exogenous: if no (sufficient) sanction is administered, the behaviour will extinguish. We will get back to these effects of transmission mechanisms in the following subsection.

The strength with which beliefs are held obviously affects their transmission: the stronger the belief, the higher the probability that it will be transmitted to others. In addition, the stronger the belief, the stronger will be the impact it will have on the recipient (cf. Latané 1981). However, this does not concern all beliefs in the same way: the propagation of rumours and gossip, especially those concerning the reputation of other agents or categories of agents, is relatively independent of the beliefs' objective truth value and of the strength with which agents hold them. The success of these beliefs is based on the motives of their acceptance (prudence) and on the mechanism of transmission itself. In other words, these beliefs are successful because they spread easily and quickly. And they spread easily and quickly because they represent a sort of 'reciprocal altruism of beliefs'. Simulation data (Castelfranchi et al. 1998; Saam and Harrer 1999) provide good evidence of this mechanism applied to the spread of social norms, or to collective misbeliefs (false beliefs, cf. Doran 1998). In artificial (as well as in natural) societies, conditions are such that agents which respect the norms will be out-competed by those which violate the norm because observers will obtain much lower pay-offs than violators. Norm-abiding action is self-defeating unless information about the identity of the cheaters circulates among the good guys, who will then punish the cheaters. Indeed, the faster the spread of transmission, the higher will be the payoffs to the honest (Paolucci et al. 1999).

This phenomenon can be supposed to play a crucial role in large societies, where repeated encounters, and therefore direct retaliation of the good guys, are quite unlikely. Vampire bats (cf. Dawkins 1976) exhibit reciprocal altruism within small groups (sharing a cavern) where a benefit can be returned by the fellow which received it. But how to explain altruistic, cooperative or norm-abiding behaviour in large groups? Simulation data suggest the hypothesis that gossip, a special case of the 'reciprocal altruism of beliefs', rescues the good guys.

The specific features of this mechanism deserve attention. First, the reciprocation of beliefs is a cheap form of reciprocation (it costs but a communicative act). Second, acceptance is likely because it is essentially prudential (does not require certitude). Third, a special form of acceptance occurs: acceptance without responsibility. Gossip works as an impersonal source: agents can pass it to others without taking responsibility ('I am told that . . .'). Fourth, the mechanism is quite useful: it permits individuals to spare the costs of direct acquaintance. As a consequence, the process is highly efficient and uncontrollable. Once a piece of gossip starts to spread, it will certainly take effect. The good question is what is the range of influence of this as well as other forms of reciprocal altruism of beliefs. Which types of beliefs may it concern and what are the domains of application? For example, gossip may be expected to play a role in the spread of social prejudices, intolerance, and discrimination, especially since most of these phenomena refer to hypothetical socially dangerous categories of agents and require only prudent acceptance. In social contexts where there is competition about scarce resources, information transmission relative to the resources is obviously more costly, and belief acceptance is more conservative. How would the reciprocation of beliefs work in such a context? A careful analysis and simulation studies of these situations might help.

Normative influence is another mechanism of memetic transmission. Social norms have a great memetic impact, since they not only spread under the action of institutional forces but also spontaneously and gradually thanks to social influence. Normative influence conveys a special type of meme (i.e., a norm), which agents might accept and therefore transmit to others. Moreover, normative influence is a very fertile memetic process. As soon as something is perceived as a norm, the probability that it will spread through the population is a function of at least two combined factors (Conte and Castelfranchi 1999): its mandatory strength increases the probability that the norm will be executed, and therefore its fertility, since other agents will infer it from behaviour; in turn, the execution of the norm leads the good guys to influence others, subject to the same norms, to do the same (which is what some authors, e.g., Heckathorn 1990; Macy and Flache 1995, call social control). Normative influence not only enforces the norms, but strengthens the memetic effect: it allows the spread of a norm through behaviour and social control.

To what extent memes spread: Hypotheses about transmissibility

The paragraphs below includes examples of the social propagation of behaviour. Some of the examples (the first five) do not imply the transmission of memes while the remaining seven do. These examples illustrate that a social cognitive analysis allows us to make hypotheses about their transmissibility. In particular, behavioural propagation 'without memes' is expected to be faster and less durable, whereas propagation 'with memes' proceeds more slowly but has a deeper and longer impact. In these phenomena, behaviour does not spread automatically, but through the agents' minds. It is a deeper type of influence, and the deeper the impact, the longer it can be expected to last.

This social cognitive analysis also allows us to make more specific hypotheses about the transmissibility of memetic processes: different motives of acceptance and mechanisms of memetic transmission justify different expectations about their transmissibility.

1. *'Black-out' effect*, or restriction of the space of possible actions. Thanks to a severe restriction of feasible actions, agents converge on the same behaviour (think of the explosion of the birth rate nine months after a black-out). Here, no meme is spreading. The high regularity, or convergence, in agents' behaviours is due to some central extraordinary event. No mutual influence is exercised by the agents undergoing this effect. No meme circulates in the social space.

2. *The 'party-shower' effect*.[3] After the 1997–98 repeated earthquakes in Central Italy, people were reported to develop compulsory paranoid thoughts. As in the black-out effect, a major discontinuity had been introduced in their normal lives by a non-ordinary event. But unlike the previous effect, in this case, the influence of this event on agents is determined by their perception and interpretation of the event, and by the consequent feeling of powerlessness. However, no memetic process is (necessarily) at stake: agents did not need to communicate these feelings to one another (although, in fact, they most certainly did) for it to spread through the whole group.

3. *Behavioural 'domino' effect*. Consider the case in which, in social or public settings (e.g., a crowded restaurant), you are obliged to raise

---

[3] This name is after Searle's (1990) example of the prompt flight of participants at an outdoor party at the first raindrops of an incipient shower.

your voice so that your friends can hear you. Here, no memetic effect occurs, since agents do not form any representation of the effect they spread and contribute to amplify. They simply raise their voice in order to be listened to, thereby causing the overall level of noise[4] to increase (within a threshold, beyond which communication is unfeasible). In this phenomenon, the behavioural convergence is an indirect effect of agents' behaviour on, and through, one another.

4. *Automatic contagion of emotion expression.* Social transmission of the behavioural expression of emotions may be purely automatic (i.e., need not imply any memetic process). Think of the spread of the behavioural expression of emotions, which abound in everyday life (Freedman and Perlick 1979). This actually falls in the wide and generic category of behavioural contagion, which has been explained in terms of two different mechanisms (see Marsden, 1998): social learning and 'social release' (Ritter and Holmes 1969; Wheeler 1966; Levy and Nail 1993; for a recent analysis, see again Marsden 1998). Social release essentially consists of a mechanism by means of which, in the presence of others, individuals release behaviours that belong to their repertoire which were previously inhibited. Both groups of theories, indeed, fail to capture the main difference between contagion and other processes of propagation: the social learning theories do not account for any such difference; the social release theories reduce this difference to a strictly behavioural difference: a behaviour which spreads through contagion is already in one's repertoire, whilst a learned behaviour does not yet belong to one's repertoire. Finally, social contagion is sometimes meant in the rather broad sense of social propagation (Reber, 1995; Marshall, 1994). For example, it is unclear what is meant by 'suicidal contagion' (e.g., Phillips 1974). The spread of suicide is a rather complex phenomenon which may be due to several mechanisms including but not reduced to contagion.

5. *The 'vulnerable position' effect.* On the highway, if everybody exceeds the speed limit, you are obliged to do the same (break the norms) in order not be hit sooner or later from behind. Your behaviour is

---

[4] This is also known as the 'arena' effect: if during the performance, people in the first rows stand up, those who are right behind are automatically induced to follow their behaviour, and so on, until those people occupying the farthest seats.

influenced by the frequential norm established by others. Here, agents' mutual influence is determined by each one figuring out the consequence of diverging from a perceived regularity. However, no meme is spreading: agents do not update their representation of (a given subset of) norms.

6. *Emotional sharing.* Think of what social psychologists call *empathy* (cf. Hoffman 1975). In this type of phenomenon, memes do spread, although not in an identical format. Consider the beggar case: while he shows helplessness and even despair because he believes 'How dreadful: *I am* helpless', the empathic passer-by will feel sad because he believes 'How dreadful: *he is* helpless'. Thanks to the empathic mechanism, the passer-by shares (to some extent and for a short time) the emotion or feeling expressed by the beggar. Here, something new occurs: the passer-by perceives the emotional state of the beggar and infers his more general (social) state. Empathy is in fact based on specified attributions: people do not share the feelings of those who are perceived as responsible for their mishaps. On given attributions, they may share the feelings of the victim. The emotional sharing is therefore caused by an inferential process, by a reasoning applied to the mental and objective conditions of the victim. However, no indirect influence occurs yet.

7. *Socially based generation of beliefs.* But what happens if the sight of a helpless beggar engenders pessimistic speculations? The witness might start to reason about the cruelties of life. She may even gradually develop a negative mood (which is not only empathic, but more general and far-reaching) as a consequence of her negative perspective on life. Interestingly, such speculations are not intentionally induced by the beggar, whose implicit goal was at most to engender empathy. Negative evaluations are generated by the passer-by, but they take a social perception as an input. A share of the phenomenon of suicide waves might be explained analogously.[5]

8. *Socially based goal-activation.* This is certainly one of the most effective and recurrent forms of memetic influence. Agents infer necessities or goals from others' behaviours. It is an interesting form

---

[5] One might wonder whether the frequency of the input is or is not a memetic phenomenon. Now, while the occurrence of inputs like beggars is mostly determined by non-memetic, and even non-exclusively social factors, it is perhaps more questionable whether the frequency of suicide is already a memetic phenomenon. Certainly, the diffusion of a (literary) suicidal fashion is. However, a sudden increase of the suicide rate might simply be explained as a 'party-shower' effect.

of social facilitation: agents' inference may activate their corresponding goals, and only as an effect of this activation, they may decide to exhibit the input's behaviours (by more or less faithfully copying, or by simply resorting to a shared knowledge base). Consider a famous example from Max Weber: suppose that in the street, you see someone opening his umbrella. You almost certainly infer that it is raining, although your thick hair or wide hat prevented you to perceive the first drops. This inference will activate a goal of yours (i.e., not to get wet). Once such a goal has been activated, the role of the input agent stops. You are able to find a solution on your own. If you have an umbrella (which is already stored in your knowledge base as a good means to avoid getting wet), you will probably follow the example of your neighbour. But if you were not so mindful as to get one, you may decide to hasten your pace, or stop at the next pastry shop, or finally change your mind and retrace your steps. In all these cases, your decisions are influenced by your interpretation of the perceived passenger, but only in the former do you actually replicate the external meme (opening the umbrella). A more interesting but less neat example of this phenomenon is provided by your checking others' behaviours in order to infer whether a given norm applies and must be followed or not: 'A no-smoking sign is visible but everybody is smoking: smoking must be somehow tolerated . . .'

9. *Socially based value activation.* For example, I may join my colleagues who send money to the Kosovars; or follow northern Europeans who devote a part of their time to voluntary assistance, and so on. This is not yet conformity, because conformity (cf. Dignum and Conte 1997) implies a 'relativized' goal (Cohen and Levesque 1990)—that is, a goal which exists if and only if a given belief exists, and which is abandoned if the corresponding belief is revised or removed: x does action a as long as x believes that y does a and x wants to be like y. It is rather norm-based social monitoring: the activated goal is not merely relativized to conformity: x's goal is elicited by others' behaviours but can survive them. This type of goal will not be dropped, just because x perceives that others have changed their minds.

10. *The 'auction' effect.* Here, the goal of replicating others' behaviour is relativized to one's belief about others' beliefs. In the classical form of auction, agents are exposed to and are influenced by all

others' evaluations of a given commodity. They offer different eval-uations than they do in private evaluations of the same commodity (cf. Camerer and Ho in press).

11. *Elite-oriented conformity*. In this case, agents are ruled by their goal to show the same taste and preferences as those shown by (signif-icant) others. They will exhibit given tastes and standards provided that they believe that these are shared by their models. Interestingly, this is complementary to the *Simmel effect*, shown by agents who consider themselves as 'élites.' These have the goal of maintaining preferences as long as these are shared only by their affiliates. As soon as others converge on the same preferences, in order to be perceived as affiliates to the elite, the elitarian agents will drop them and turn to other, more selective, ones, and the process will be re-initialized.

12. *Norm recognition and acceptance*. While perceiving and selecting among external inputs, agents may find cues of candidate norms (cf. Conte *et al.* 1998), which they check against several tests (costs? wanted prescriptions? existing authority? etc.) before accepting them as norms.

These phenomena can be compared on a number of observational criteria, essentially drawing upon Dawkins's principles of transmissi-bility:

- *Fidelity* (or exact reproduction). The phenomena in the first six columns of Table 5.1 tend to be more regular, or exhibit lower vari-ance, than is the case in the second group. This is because in the former case the influence is direct and not diluted by indirect trans-mission, and therefore the chance of misperception decreases. At the same time, influence in these first cases is automatic and does not undergo cognitive-processing, selection, and re-elaboration.

- *Fecundity* (or indirect influence). This, of course, has an impact on the range of influence of a given phenomenon. When influence is transmitted from one agent to another, the range is higher. Usually, non-transmissible influence is confined within the range of influence of a central event. Indeed, in most of the above phenomena, influ-ence is transmissible. Agents play a twofold role: as both the patient and agent of influence, she receives and exercises it. Obviously, this amplifies the process and extends the limits of replication.

Table 5.1 Behavioural propagation: A comparison among examples

	Black-out	Party-shower	Domino effect	Contagion	Vulnerable position	Suicide effect	Empathy	Weber example	Kosovar example	Auction effect	Simmel effect	Norm recognition
Fidelity	+	+	+	+	+	+	-	-	-	-	-	-
Fecundity	-	-	+	+	+	+	+	+	+	+	+	+
Stability/ Durability	-	-	-	-	-	-	-	-	+	+	+	+
Adjustability	-	-	-	-	+	+	+	+	+	+	+	+

- *Stability* (*or durability*), that is, how long the influence lasts over time.
- *Adjustability*. This is not meant to be the dual of fidelity, but rather as the agents' accepting and adjusting the input received to their own (current) problem-solving and planning. Of course, there may be a trade-off between this feature and fidelity (but this appears to be not always the case): the former might cause lower fidelity in transmission. Presumably, behavioural transmission is a delicate balance between these two complementary aspects: fidelity in replication, and agent-oriented acquisition.

Five groups of phenomena can be identified in Table 5.1, according to the evaluation of each example on all the dimensions considered. These groups indicate at least as many types of behavioural propagation, from higher to lower fidelity, higher to lower stability, negative to positive fecundity, and negative to positive adjustability. It would be interesting to have data (perhaps simulation data) to perform a trend analysis of the correlations among these features.

Other dimensions can be discovered, and a more analytical picture might emerge. For example, one might compare these examples (or others) in terms of speed of transmission, or more precisely, speed of appearance and extinction. One might reasonably conjecture that these features correlate negatively: the faster a given phenomenon spreads over a population, the more rapidly it will decay. This conjecture seems supported by the argument that the phenomena suddenly appearing are those that imply none or poor *modification of the mind* (either permanent or temporary) of the agents: behavioural contagion, for example, is not controlled by mental processes but spreads automatically. In principle, automatic processes are quicker than controlled ones, and should spread faster. But they fade as quickly as they appear: as soon as the exposure to contagion ceases, its effects are extinguished.

This points to another interesting criterion: contingency versus autonomy of the effect. Contingent effects are those which cease with the extinguishing of their causes. Autonomous ones survive their causes, although they might then perish over time. A special case of this criterion is vertical transmission: obviously, only autonomous effects are likely to be conveyed to subsequent generations. Contingent effects can only spread horizontally.

In Table 5.2, the new criteria introduced have intermingled the groups previously identified. However, a perfect complementarity holds

Table 5.2 Behavioural propagation: A more analytical comparison

	Black-out	Party-shower	Domino effect	Contagion	Vulnerable position	Suicide effect	Empathy effect	Weber example	Kosovar example	Auction effect	Simmel effect	Norm recognition
Fidelity	+	+	+	+	+	+	–		–	–	–	
Fecundity	–	–	+	+	+	+	+	+	+	+	+	+
Stability/ Durability	–	–	–	–	+	–	–		+	+	+	+
Adjustability	–	+	–	+	+	+	+	+	–	+	+	
Speed transmission	+	+	+	+	+	±	+	±		±		
Speed extinction	+	–	+	+	+	±	+	±		±		
Contingency	+	+	+	+	+	±		–		+	+	–
Verticality	–	–	–	–	–	±	–	–	+	–	–	+

between the first and last columns; but this complementarity gradually decreases and vanishes in the central columns. Behavioural propagation may or may not imply memetic transmission, and this has effects on observable features of the propagation process: memetic processes (right-hand side of the Table 5.2) are expected to show higher stability, more autonomy than non-memetic ones, and to decline more gracefully. At the same time they are expected to present lower fidelity. Non-memetic processes (left-hand side of the Table 5.2), instead, are less autonomous, and less stable, but spread more rapidly and show higher fidelity.

But which factors or aspects of the phenomena in question allow this comparison to be made? The answer can be found in an analysis of the mental processes involved: *examples are interpreted as more stable, more autonomous—in a word as memetic—when the transmission implies that each agent influences (i.e., causes a modification of) another agent' s mind*, and *when such an influence implies the social and cognitive competencies of the agents involved (both the influencing and the influenced agents).* One can expect that the more the transmission relies on the mental modification of the agents involved, the slower the transmission and the lesser its degree of fidelity, but also the more stable and autonomous (in the sense previously defined) the behavioural effect.

Which memes spread?

Memetic processes may sometimes interefere with each other. In particular, they may be either concurrent or cooperative. Mental and cognitive processes allow these potential interferences to be detected, and possibly outputs to be predicted.

For example, social and legal norms may interfere positively and negatively with their memetic effects. Sometimes, social norms may contrast with legal norms. Although liable to influence, social agents may violate the norms, both the social and the legal ones. Often, norm violation is the result of conflicts among normative systems (e.g., between legal and social norms). Also, violation allows a solution of the norm conflict. In any case, it is impossible to explain and predict the outputs of this interference or competition among different memes (norms), without understanding why and how agents select them. To see why, consider the following example.

Suppose that, in the daylight, a car proceeding on the opposite side of the street flashes while approaching you. You are exceeding the speed limit. There are several possible interpretations here, which might produce different memetic effects. If you interpret the flashing as a *greeting*, you may return it or not, but the probability that you will replicate the same behaviour with other cars, which you may encounter afterwards, is not very high (poor transmissibility). And, suppose that some minutes after receiving the flashing, you realize that an automatic speed limit controller is situated on the highway. In these conditions, it is possible that you reinterpret the previous driver's behaviour as a *warning* (informing you that you will undergo an automatic speed limit control). If this is the case, the probability that you will replicate the same behaviour to the benefit of others (flashing to warn other drivers proceeding in the opposite direction that they are about to undergo a speed limit control) increases accordingly. To the extent that this interpretation spreads and becomes stable, drivers will keep violating the speed limits: the diffusion of flashing as a warning acts as a norm-concurrent or antagonistic memetic process.

Finally, suppose that you find no automatic control. If the first car was proceeding at a regular speed, the probability that you will interpret the flashing received as a *blame* is higher than the probability that you will interpret it as a warning. If you take it as a blame, you may decide to lower your speed, and, if you do so, you will most certainly use the same behaviour (flashing) to reproach other drivers which are proceeding at an irregular speed. To the extent that this behaviour spreads, it enforces the norm (norm-cooperative memetic process), and may contribute to a global reduction of speed.

But when do these different interpretations and their effects occur? Under which conditions, does social control enforce legal norms, and when, instead, coevolving social norms neutralize them? This is a fascinating and open question. However, the analysis proposed so far suggests hypotheses that could be tested by means of simulation. For example, when the reciprocal altruism of beliefs (in our example, flashing as a warning) is sufficient to spare both the costs of obedience (in the preceding example, speed reduction) and those of transgression (fine), the norm-antagonistic process may be expected to out-compete the norm-cooperative one. But when this is not the case—that is, when the external conditions cause the reciprocal altruism of beliefs to have costs not lower than those of transgression (transmission of beliefs is costly, dangerous, or punished)—the norm-concurrent process is bound to extinguish soon, whilst there is no reason to have the same expectation concerning the norm-cooperative process: agents that *do* comply with the norm are likely to exercise social control on others to the benefit of the norm (i.e., flashing as a blame).

The effects of memes on social behaviour

A decade or so ago, enthusiastic observers of the spread of electronic communication welcomed the Internet as the symbol of a new age of participation and cooperation, in which electronic 'social fields' emerge and 'group processes' are facilitated (*Communications of the ACM* 1994). The reasoning was apparently quite simple: since electronic means facilitate communication which is essential for participation and cooperation, then the Internet can be expected to promote participation and cooperation—for example, through non-profit communication, and civic networks. More than incomplete, or based on insufficient elements, the reasoning was wrong, as current evidence shows: the Internet has indeed spread worldwide, but its diffusion—far from promoting communitarian associations (which did not grow much after an initial spurt)—is mainly used for economic transactions, in electronic commerce.

Was this effect predictable? To some extent it was, without necessarily bringing about the argument that the Western societies are profit-oriented. Of course, they are. But the reason why the Internet could not be expected to invert this trend resides in the ingredients of cooperation versus exchange. Cooperation is a social action that requires at least two minimal conditions: that cooperating agents have one common goal and that they are interdependent to its achievement (cf. Conte and Castelfranchi 1995). Conversely, in exchange (cf. Homans 1974), agents need only to be interdependent. The probability that the conditions for cooperation apply even in a wide set of agents is at least twice as low than the probability that the condition for exchange applies. To this, other more sophisticated social cognitive factors should be added: unlike exchange, cooperation implies a common plan and a complex process of agreement (cf. Cohen and Levesque 1991). Besides, cooperation rules out or reduces the convenience of cheating (for an analysis of this point, cf. Conte and

Castelfranchi 1995), which abound in social and economic exchanges. It is therefore no surprise that the market-oriented application of the Internet has had a much wider influence than its cooperative use. Nonetheless, the hope for more cooperative applications of information technologies should not be abandoned. I will turn to this point in the following section.

Further advantages: Memetics for MAS

Information societies are hybrid multi-agent systems where human agents coexist and interact with software agents. One typical example is agent-mediated electronic commerce. So far, in this context, software agents have been mainly used to find the best bargain (Bargainfinder, http://bf.cstar.ac.com). Essentially, these agents search the Internet in an intelligent way (see Doorenbos *et al.* 1996). Other applications consist of electronic marketplaces where agents sell and buy goods (a good example is KASBAH; Chavez and Maes 1996).

But these applications of software agents are also insufficient because they are too competitive. Indeed, Bargainfinder 'was one early agent that managed to get banned from a number of CD (compact disk) stores because its aims were possibly not beneficial to any particular store' (Crabtree 1998: 135). A mediator or representative agent must be accepted by the community into which it will interact. Hence, agent systems for negotiation must be enabled to deal with this problem (Gutmann *et al.* 1998). In order to act in the interests of their users, software agents must be able to detect errors and irregularities in contracts, to negotiate with partners (and not only to find them). But they must do so without providing private information about their users, and without breaking other social conventions. In sum, software agents are expected to be norm-abiding and cooperative even in a competitive context, like the market.

Which properties enable agent systems to accept useful social laws or conventions (such as respect privacy)? How to make them avoid socially undesirable behaviour (e.g., do not cheat)? This is far from simple. Implementing mere constraints in the agent's action repertoire is insufficient: agents ought to be able to choose whether to resist external influence (do not deliver information if this is dangerous for the user), or accept it (accept and use information about others' reputation), and determine whether to lie (about reserved price, or private information) or not (do not cheat if this diminishes your client's reputation). Briefly, future software intermediaries for electronic trans-actions must be memetic agents endowed with the capacity to select and accept beliefs and transmit them.

A cognitive variant of a memetic glossary

Let us clarify the way some expressions, typically used by memeticists, should be rephrased in cognitive terms:

- *Meme.* In this presentation, a meme is meant as a symbolic repre-sentation of any state of affairs. In this sense, memes are both internal, implemented in the mind, or external, for example, incorporated or implemented in an external (non-mental) object.

- *External memes.* These are directly accessible objects (artefacts, prod-ucts, behaviours) that incorporate a meme. External implementation of a meme is the activity implied in the production of objects and performance of behaviours which incorporate memes. *Caveat*: the fact that an object is directly accessible does not make the memes it conveys or incorporates easy to decode.

- *Internal memes (or mental implementation).* This is rather complex. A meme's internal implementation is a process which memeticists do not seem to perceive (cf. Rhodes 1999). As I said above, memes are usually equated with concepts and memetic agents with *recipi-ents* and *vectors* of memes.[6] The memetic process is then seen as the storing (and selecting) of memes and their associated values of impor-tance. This simplifies the memetic process to the point of rendering it opaque. Four aspects of the memetic process are thereby ignored: (1) *belief- and goal-generation*: the agents' desire to form new repre-sentations (beliefs and goals), and their resorting to other agents to acquire them; (2) *belief- and goal-adoption*: the agents' decision to accept external representations, and the mechanisms allowing them to select among candidate ones; (3) *integration* of candidate repre-sentations with the internal ones (which should not be seen as a list, again, see Rhodes 1999); (4) *implementation* of this internal

---

[6] The terminology recipient/vector is preferred here over that of receiver/sender to clarify that memetic transmission is not necessarily a communicative process.

representation into an externally accessible phenomenon (behaviour, product). Interestingly, this external effect may include emotional expression.

- *Memetic process.* This is a process by means of which memes replicate. In particular, memes replicate memetically (as opposed to, say, epidemically) when they propagate: (a) through the *social* minds of the agents, thanks to their social competence, and (b) across their minds, meaning from one to another. In order for a meme to propagate memetically, it must undergo the mental process described earlier: autonomous agents must be social enough as to able to need and implement representations, resort to others, perceive external candidate representations, filter them according to their internal criteria, and reimplement them into their behaviours, thereby contributing to the replication of the meme.

- *Memetic agent.* This is a limited *autonomous social* agent, endowed with social competence.

- *Social competence.* This includes but is not limited to imitation and language (see above). Interestingly, a memetic agent may have a specific social competence, provided or required by his or her role (see Wilkins 1998). But role-playing is not the only social competence required for propagating memes. A more fundamental level of sociality, which implies the capacity to detect and reason upon others' mental states, is needed. Any memetic agent is a social agent. Any memetic agent is a recipient and a vector of memetic processes. However, a social agent does not necessarily act memetically. For example, he or she may select out some candidate external meme. On the other hand, together with role-adoption and role-playing, social competence includes the capacity to detect and represent prescriptive expectations, such as social norms.

Conclusion

In this chapter, the field of memetics is observed from a rather specific perspective—the study of social cognitive processes among limited autonomous agents. These processes are argued here to be essential in an account of cultural change and evolution, and more specifically in memetic processes.

After a brief reconsideration of the (many) advantages and (some) disadvantages of the field, one main aspect of memetic theory is found unsatisfactory. This was the treatment of the memetic agent, and the conceptualization of the requirements of memetic processes. The work presented in this chapter is then concentrated on this issue.

A model of a limited autonomous agent is briefly presented, which defines the social agent as liable to social influence but at the same time endowed with internal criteria and motives for accepting it. Next, this model is shown to be able to account for investigating the mechanisms of memetic transmission, making hypotheses about their transmissibility, observing and predicting the effects of memetic processes.

Essentially, memetic agents are argued to be limited autonomous agents endowed with a capacity for social action.

Of course, the hypotheses discussed in this chapter are rather preliminary and would certainly profit from a more accurate analysis of idealized examples of memetic and non-memetic transmission, and from a more systematic investigation of criteria for comparison. How to determine the effects of social competence on the nature and characteristics of social and cultural transmission? How to control the hypothesis that specific mental processes are responsible for specific observable features of behavioural and/or cultural change and evolution? A suitable research method is offered by the field of social simulation and artificial societies. Some memeticists are familiar with the techniques and languages of social simulation (cf. the El Farol Bar model, Edmonds 1998). However, a closer interplay between these fields is desirable, on the grounds of the promising subfield of *agent*-based social simulation. This would certainly allow memetics to actualize its theoretical potential and invest in the exploration of some well-defined phenomena. It would also contribute to scientific recombination and innovation: the field of (agent-based) SiMetics (simulated Memetics) is not too bad as a meta-memetic effect—that is, as an outcome of a memetic process about memetics!

Acknowledgements

I would like to thank my colleagues from IP-CNR Cristiano Castelfranchi, Rino Falcone, Maria Miceli, and Rainer Hegselmann from the University of Bayreuth, Germany, for their valuable comments on previous drafts of this text.

References

Bandura, A. (1971). *Social learning theory.* New York: General Learning Press.

Benvenuto, S. (2000). *Dicerie e pettegolezzi.* Bologna: Il Mulino.

Blackmore, S. (1999). *The meme machine.* Oxford: Oxford University Press.

Camerer, C. F. and Ho, T. H. (in press). Experience-weighted attraction learning in normal-form games. *Econometrica.*

Castelfranchi, C. (1997). Principles of limited autonomy. In *Contemporary Action Theory* (ed. R. Tuomela and G. Holmstrom-Hintikka), Dordrecht: Kluwer, pp. 315–45.

Castelfranchi, C., Conte, R., and Paolucci, M. (1998). Normative reputation and the costs of compliance. *Journal of Artificial Societies and Social Simulation,* 1(3). [http://www.soc.surrey.ac.uk/JASSS/1/3/3.html]

Castelfranchi, C., Treur, J., Dignum, F., and Jonker, C. (1999). A BDI architecture for normative agents. *Proceedings of Agent Theory, Architecture and Language* (ATAL 99) Berlin: Springer, pp. 209–27.

Cavalli-Sforza, L. L. and Feldman, M. (1981). *Cultural transmission and evolution. A quantitative approach.* Princeton: Princeton University Press.

Cecconi, F. and Parisi, D. (1998). Individual versus social survival strategies. *Journal of Artificial Societies and Social Simulation,* 1(2). [http://www.soc.surrey.ac.uk/JASSS/1/2/1.html]

Chavez, A. and Maes, P. (1996). Kasbah: An agent marketplace for buying and selling goods. In the *First International Conference On the Practical Application of Intelligent Agents and Multi-Agent Technology,* London pp. 75–90.

Cohen, P. R. and Levesque, H. J. (1990). Persistence, intention, and commitment. In *Intentions in Communication* (ed. P. R Cohen, J. Morgan, and M.A. Pollack), pp. 33–71. Cambridge, MA: MIT Press.

Cohen, P. R. and Levesque, H. J. (1991). *Teamwork.* Technical Report, SRI-International, Menlo Park, CA.

Conte, R. (1999). Social intelligence among autonomous agents. *Computational and Mathematical Organization Theory* 5: 203–29.

Conte, R. and Castelfrachi, C. (1995). *Cognitive and social action.* London: UCL Press.

Conte, R. and Castelfranchi, C. (1999). From conventions to prescritions. Towards an integrated view of norms. *Artificial Intelligence and Law,* 7: 323–40.

Conte, R. and Dignum, F. (forthcoming). From social monitoring to normative influence. Paper presented at the *International Meeting on Modelling Agent Interactions in Natural Resource and Environment Management,* INRA ENSAM Campus, Montpellier, France.

Conte, R., Hegselmann, R., and Terna, P. (ed.) (1997). *Simulating social phenomena.* Berlin: Springer.

Conte, R., Castelfranchi, C. and Dignum, F. (1998, July). Autonomous norm-acceptance. In *Proceedings of Agent Theory, Architecture and Language* (ATAL 98). Paris, La Villette Berlin: Springer, pp. 48–64.

Crabtree, B. (1998). What chance software agents. *The Knowledge Engineering Review,* 13: 131–7.

Dawkins, R. (1976). *The selfish gene.* Oxford: Oxford University Press.

de Jong, M. (1999). Survival of the institutionally fittest concepts. *Journal of Memetics–Evolutionary Models of Information Transmission*, 3. [http://www.cpm.mmu.ac.uk/jom-emit/1999/vol3/de_jong_m.html]

Dennett, D. (1995). *Darwin's dangerous idea*, London: Allen Lane Press.

Dignum, F. and Conte, R. (1997). Intentional agents and goal formation. In *Proceedings of the 4th International Workshop on Agent Theories Architectures and Languages* (ed. M. Singh *et al.*), Providence: Springer, pp. 118–32.

Donald, M. (1991). *Origins of the modern mind*. Cambridge, MA: Harvard University Press.

Doorenbos, B., Etzioni, O. and Weld, D. (1996). *A scalable comparison-shopping agent for the World Wide Web*. Technical Report, TR96–01–03, Washington, DC: University of Washington.

Doran, J. (1994). Modelling collective belief and misbelief. In *AI and cognitive science '94* (ed. M. Keane, *et al.*), pp. 89–102, Dublin University Press.

Doran, J. (1998). Simulating collective misbelief. *Journal of Artificial Societies and Social Simulation*, 1(1). [http://www.soc.surrey.ac.uk/JASSS/1/1/3.html]

Edmonds, B. (1998). On Modelling in Memetics. *Journal of Memetics–Evolutionary Models of Information Transmission*, 2. [http://www.cpm.mmu.ac.uk/jom-emit/1998/vol2/edmonds_b.html]

Frank, J. (1999). Applying memetics to financial markets: Do markets evolve towards efficiency? *Journal of Memetics–Evolutionary Models of Information Transmission*, 3. [http://www.cpm.mmu.ac.uk/jom-emit/1999/vol3/frank_j.html]

Freedman, J. L., and Perlick, D. (1979). Crowding, contagion and laughter. *Journal of Experimental Psychology*, 15: 295–303.

Gabora, L. (1997). The origin and evolution of culture and creativity. *Journal of Memetics–Evolutionary Models of Information Transmission*, 1. [http://www.cpm.mmu.ac.uk/jom-emit/vol1/gabora_l.html]

Gilbert, N. and Conte, R. (ed.). (1995). *Artificial societies: The computer simulation of social life*. London: UCL Press.

Gilbert, N. and Doran, J. (ed.). (1994). *Simulating societies: The computer simulation of social processes*. London: UCL Press.

Gilbert, N. and Troitzsch, K. (1999). *Simulation for social scientists*. Milton Keynes: The Open University.

Gutmann, R. H., Moukas, A. G. and Maes, P. (1998). Agent-mediated electric commerce: A survey, *The Knowledge Engineering Review*, 13: 147–61.

Heckathorn, D. D. (1990). Collective sanctions and compliance norms: a formal theory of group-mediated social control. *American Sociological Review*, 55: 366–83.

Hoffman, M.L. (1975). Altruistic behaviour an the parent-child relationship. *Journal of Personality and Social Psychology*, 31: 937–43.

Homans, G. C. (1974). *Social behaviour. Its elementary forms*. New York: Harcourt.

Latané, B. (1981). The psychology of social impact. *American Psychologist*, 36: 343–56.

Leslie, A. (1992). Pretense, autism and the theory-of-mind module. *Current Directions in Psychological Science*, 1: 18–21.

Levy, D. A. and Nail, P. R. (1993). Contagion: A theoretical and empirical review and reconceptualization. *Genetic, Social and General Psychology Monographs*, 119: 235–183.

Macy, M. and Flache, A. (1995). Beyond rationality in models of choice. *Annual Review of Sociology*, **21**: 73–91.

Markus, H. and Zajonc, R. B. (1985). The cognitive perspective in social psychology. In *Handbook of Social Psychology* (ed. G. Lindzey and E. Aronson) Hillsdale, NJ: Erlbaum.

Marsden, P. (1998). Memetics and social contagion: Two sides of the same coin? *Journal of Memetics–Evolutionary Models of Information Transmission*, **2**. [http://www.cpm.mmu.ac.uk/jom-emit/1998/vol2/marsden_p.html]

Marshall, G. (ed.) (1994). *Concise oxford dictionary of sociology*. Oxford: Oxford University Press.

Paolucci, M., Marsero, M., and Conte, R. (1999). What's the use of gossip? A sensitivity analysis of the spread of respectful reputation. In *Tools and techniques for social science simulation* (ed. R. Suleiman, K. G. Troitzsch, and G. N. Gilbert), Heidelberg: Physica, pp. 302–17.

Phillips, D. P. (1974). The influence of suggestion on suicide: Substantive and theoretical implications of the Werther effect. *American Sociological Review*, **39**: 340–54.

Preti, A. and Miotto, P. (1997). Creativity, evolution and mental illnesses. *Journal of Memetics–Evolutionary Models of Information Transmission*, **1**. [http://www.cpm.mmu.ac.uk/jom-emit/1997/vol1/preti_aandmiotto_p.html]

Rao, A. S. and Georgeff, M. P. (1991). Modelling rational agents within a BDI architecture. In *Proceedings of the International Conference on Principles of Knowledge Representation and Reasoning* (ed. J. Allen, R. Fikes, and E. Sandewall), San Mateo, CA: Kaufmann, pp. 473–85.

Reber, A. S. (ed.) (1995). *The penguin dictionary of psychology* (2nd edn). London: Penguin.

Rhodes, T. (1999). Memetic vector modeling: A quest for the mathematics of memes. [Paper available at http://www.speakeasy.org/~proftim/memes/]

Ritter, E. H. and Holmes, D. S. (1969). Behavioral contagion: Its occurrence as a function of differential restraint reduction. *Journal of Experimental Research in Personality*, **3**: 242–6.

Rockloff, M. J., and Latané, B. (1996). Simulating the social context of human choice. In *Social Science Microsimulation* (ed. K. G. Troitzsch, U. Mueller, N. Gilbert and J. Doran), Berlin: Springer, pp. 359–75.

Saam, N. J. and Harrer, A. (1999). Simulating norms, social inequality, and functional change in artificial societies. *Journal of Artificial Societies and Social Simulation*, **2**(1). [http://www.soc.surrey.ac.uk/JASSS/2/1/2.html]

Searle, J. (1995). *The construction of social reality*. London: Penguin.

Sherif, M. (1936). *The psychology of social norms*. New York: Harper & Row.

Sichman, J. S., Conte, R., Castelfranchi, C., and Demazeau, Y. (1994). A social reasoning mechanism based on dependence networks. In *Proceedings of the 11th European Conference on Artificial Intelligence* (ed. A. G. Cohn), pp. 188–92. Chichester: Wiley.

Sichman, J. S., Conte, R., and Gilbert, N. (ed.) (1998). *Multi-agent systems and agent-based simulation*. Berlin: Springer.

Sierra, C. (forthcoming). *Agent-mediated electronic commerce: A European perspective*, Springer.

Simon, H. A. (1969). *The sciences of the artificial*, Cambridge, MA: MIT Press.

Weiss, G. (ed.) (1999). *Multiagent systems: A modern approach to distributed artificial intelligence*, Cambridge, MA: MIT Press.

Wheeler, L. (1966). Towards a theory of behavioural contagion. *Psychological Review*, 73: 179–92.

Wilkins, J. S. (1998). What's in a meme? Reflections from the perspective of the history and philosophy of evolutionary biology. *Journal of Memetics–Evolutionary Models of Information Transmission*, 2. [http://www.cpm.mmu.ac.uk/jom-emit/1998/vol2/wilkins_js.html]

Wooldridge, M. (1999). Intelligent agents. In *Multiagent systems: A modern approach to distributed artificial intelligence* (ed. G. Weiss), pp. 27–78, Cambridge, MA: MIT Press.

CHAPTER 6

The evolution of the meme

Kevin N. Laland and John Odling-Smee

Towards an understanding of culture

In 1871, Tylor defined culture as 'that complex whole which includes knowledge, belief, art, morals, custom and any other capabilities and habits acquired by man as a member of society'. Although in anthropological circles Tylor's rather cumbersome definition has been superseded, it still captures the intuitive notion of culture held by the layperson. Moreover, it illustrates a challenge—arguably the greatest challenge—for the evolutionary biologist; namely, how could such a seemingly inextricably interwoven complex of ideas, behaviour, institutions and artefacts evolve?

In our view, biologists and human scientists alike will not be able to understand the evolution of culture unless they are prepared to break down the 'complex whole' into conceptually and analytically manageable units. To this end, we regard memes as a valuable scientific tool. We find compelling the psychological evidence for memes as packages of learned and socially transmitted information, stored as discrete units, chunked and aggregated into higher order knowledge structures, encoded as memory traces in interwoven complexes of neural tissue, and expressed in behaviour. For us, the pertinent question is not whether memes exist, as suggested by Aunger in the Introduction, but whether they are a useful theoretical expedient. In this chapter, we describe our views on the evolution of culture, and sketch how 'memes' help elucidate that story.

However, we begin with two qualifications. First, we have concocted little more than a plausible story about the evolution of culture. While we are prepared to defend our narrative, we recognize that there is a long way to go. Second, although memes are central to our views of culture, we do not believe that culture is simply a collection of memes. Yet if we are to make progress in understanding cultural change, it is likely to be useful to distinguish between the informational and non-informational components of culture, and to acknowledge the human tendency constantly to build up, break down, and rebuild ideational complexes.

In the first section we outline our evolutionary perspective, placing emphasis on the capacity of organisms to modify their environments, which we call 'niche construction' (Odling-Smee 1988). We suggest that complex organisms have evolved a set of information-gaining processes that are expressed in niche construction, and that a capacity for acquiring and transmitting memes is one such process. We go on to argue that, as many animals are capable of learning from others, they too can be said to have memes, and describe how animal protoculture might have evolved into human culture through meme-based niche construction. In the penultimate section, we use our evolutionary framework to suggest that the success of a meme depends not just on its infectiousness, but also on the susceptibility of the host, and on the social environment. Finally, we present an example from gene–culture coevolutionary theory to illustrate how a formal theory of memetics can be of value.

Niche construction

Culture has allowed human beings to change their environments dramatically. Yet humans are not alone in modifying their world. Many other species do or have done the same, mostly without any help from culture (Lewontin 1983, 2000; Odling-Smee *et al.* 1996; Jones *et al.* 1997). Elsewhere, we have argued that the significance of evolutionary theory to the human sciences cannot be fully appreciated without a more complete understanding of how phenotypes in general modify significant sources of selection in their environments (Laland *et al.* 2000).

Our understanding of the evolution of culture begins, not with the meme, but with another of Dawkins' important insights, the 'extended phenotype'. Dawkins (1982) argued that genes can express themselves outside the bodies of the organisms that carry them. For instance, the beaver's dam is an extended phenotypic effect of beaver genes, while the houses of caddis fly larvae are equally expressions of caddis fly genes. In fact, the genes of all organisms express products that impinge on the environment. A basic feature of living creatures is that they take in and assimilate materials for growth and maintenance, and eliminate or excrete toxic waste products. It follows that, merely by existing, organisms must change their local environments to at least some small degree.

At first sight it may be tempting to conclude that the impact that most organisms have on their environments is trivial, a mere drop in the ocean compared with the action of major geophysical, chemical, or meteorological processes. A closer inspection reveals that countless organisms across the breadth of all known taxonomic groups significantly modify their local environments (Lewontin 1983, 2000; Odling-Smee 1988; Jones *et al.* 1997). To varying degrees, organisms choose their own habitats, mates, and resources and construct important components of their own, and their offsprings' local environments, such as nests, holes, burrows, pupal cases, paths, webs, dams, and chemical environments. Following Lewontin (1983), we argue that organisms not only adapt to environments, but in part also construct them.

Niche construction starts to take on a new significance when it is acknowledged that, by changing their world, organisms modify many of the selection pressures to which they and their descendants are exposed, and that this may change the nature of the evolutionary process. To go back to the beaver, its dam sets up a host of selection pressures that feed back to act not only on the genes that underlie dam building, but also on other genes that may influence the expression of other traits in beavers, such as their teeth, tail, feeding behaviour, their susceptibility to predation or disease, their social system, and many other aspects of their phenotypes. Dam construction may also affect many future generations of beavers that may 'inherit' the dam, its lodge, and the altered river, as well as many other species of organisms that now have to live in a world with a lake in it. Niche construction generates a form of feedback in evolution that is not yet fully appreciated by contemporary evolutionary theory (Lewontin 1983, 2000; Odling-Smee 1988; Odling-Smee *et al.* 1996; Laland *et al.* 1996a, 1999).

There are numerous examples of organisms choosing or changing their habitats, or of constructing artefacts, leading to an evolutionary response. For instance, web spiders construct webs, which have led to the subsequent evolution of camouflage, defence, and communication behaviour on the web (Preston-Mafham and Preston-Mafham 1996). Similarly, ants, bees, wasps, and termites construct nests that are themselves the source of selection for many nest regulatory, maintenance and defence behaviour patterns (Hansell 1984; Holldobler and Wilson 1994). Countless mammals, reptiles, and amphibians construct burrow systems or nests, and here too there is evidence that behaviour underlying nest complexity, defence, maintenance, and regulation has evolved in response to selection pressures that were initiated by initial nest construction (Hansell 1984; Nowak 1991).

Of course, this will be no surprise to the biologically minded, yet the breadth and scale of niche construction will surprise many. Few people realize that there are more than 34 000 species of spider that construct silken egg sacs, burrows or webs (Preston-Mafham and Preston-Mafham 1996). There are more than 9000 species of birds, the vast majority of which construct nests (Forshaw 1998), and probably as many fish that do the same (Paxton and Eschmeyer 1998). There are 9500 known species of ants, and 2000 known species of termites, all living in social colonies, and almost all building some kind of nest (Holldobler and Wilson 1994; Gullan and Cranston 1994). Niche construction is all-pervasive.

Most cases of niche construction, however, do not involve the building of artefacts, but merely the selection or modification of habitats. For example, as a result of the accumulated effects of past generations of earthworm niche construction, present generations of earthworms inhabit radically altered environments where they are exposed to modified selection pressures (Darwin 1881; Lee 1985). Odling-Smee (1988) has described this legacy of modified selection pressures as an 'ecological inheritance'. Females of the vast majority of the millions of insect species lay eggs, and usually the eggs are deposited on or near the food required by the offspring upon hatching (Gullan and Cranston 1994). This is probably one of the most frequently documented cases of ecological inheritance. The offspring of virtually all insects inherit from their mother a legacy of readily available, nutritious larval food.

Figure 6.1 shows how niche construction and ecological inheritance interact with natural selection and genetic inheritance. Figure 6.1a represents the standard evolutionary perspective: organisms transmit genes from one generation to the next, under the direction of natural selection. Figure 6.1b extends this perspective to acknowledge that

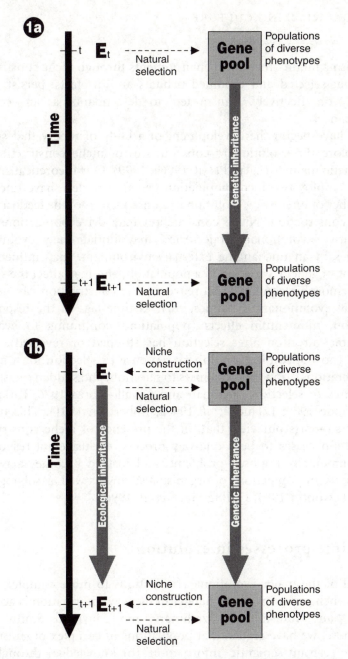

Figure 6.1. (a) The standard evolutionary perspective: populations of organisms transmit genes from one generation to the next, under the direction of natural selection. (b) With niche construction, phenotypes modify their local environments (E) through niche construction. Each generation inherits both genes and a legacy of modified selection pressures (ecological inheritance) from ancestral organisms. Reprinted by permission of Cambridge University Press from Laland, K.N., Odling-Smee, F.J. and Feldman, M.W. (2000). Niche construction, biological evolution and cultural change. *Behavioural and Brain Sciences*, **23**: 1–46.

organisms modify their local environments through niche construction, and that selected and modified habitats and artefacts, persist, or are actively or effectively 'transmitted' to descendants, as an ecological inheritance.

We have begun the development of a body of theory that sets out to explore the evolutionary consequences of niche construction in a systematic manner (Laland *et al.* 1996a, 1999). Our theoretical analyses, which employ two-locus population genetics models, have uncovered a number of interesting evolutionary consequences of the feedback from niche construction. Niche construction may drive populations along alternative evolutionary trajectories, may initiate new evolutionary episodes in an unchanging external environment, may influence the amount of genetic variation in a population, and may affect the stability of polymorphic equilibria. Moreover, niche construction can generate unusual evolutionary dynamics, such as time-lags in the response to selection, momentum effects (populations continuing to evolve in the same direction after selection has stopped or reversed), inertia effects (no noticeable evolutionary response to selection for a number of generations), opposite responses to selection, and sudden catastrophic responses to selection (Feldman and Cavalli-Sforza 1976; Kirkpatrick and Lande 1989; Laland *et al.* 1996a; Robertson 1991). This body of theory supports our view that, in the presence of niche construction, adaptation ceases to be a one-way process, exclusively a response to environmentally imposed problems, and instead becomes a two-way process, with populations of organisms setting as well as solving problems (Lewontin 1983; Odling-Smee *et al.* 1996).

Multiple processes in evolution

Several of the major evolutionary transitions to more complex organisms or behaviour involved changes in the way information is acquired, stored, and transmitted (Szathmáry and Maynard Smith 1995). Elsewhere, we have argued that populations of complex organisms can acquire relevant semantic 'information' (or knowledge) through a set of information-acquiring processes operating at three different levels (Laland *et al.* 2000). These processes are: (i) the population genetics processes of biological evolution; (ii) ontogenetic processes, such as learning and the immune system; and (iii) culture, or protoculture.

In each case, the knowledge gained is both expressed in niche construction, and selected by environments that are partly niche-constructed. In various combinations, these are the processes that supply all organisms with the knowledge that underlies their adaptations. Similar multiple-process models of evolution have been proposed elsewhere (Plotkin and Odling-Smee 1981; Dennett 1995).

As a consequence of the differential survival and reproduction of individuals with distinct genotypes, genetic evolution results in the acquisition, inheritance, and transmission of genetically encoded knowledge by individuals in populations. This genetic information both underpins niche construction, and is subject to selection from niche-constructed environments.

In addition, many species have evolved a set of more complicated ontogenetic processes that allow individual organisms to cope with types and rates of environmental change that they cannot deal with at the genetic level. These processes are products of genetic evolution, and are based on specialized information-acquiring subsystems in individual organisms, such as brain-based learning in animals, or the immune system in vertebrates. These ontogenetic processes are characterized by the capacity for additional, individually based information acquisition. However, unless the species concerned is capable of social learning, the adaptive knowledge acquired through these ontogenetic processes cannot be inherited because all the knowledge gained by individuals during their lives is erased when they die. Nonetheless, learned knowledge can guide niche construction. Moreover, the reverse is also true: niche construction can guide learning. Because environments are partly niche-constructed and each individual's learning will be shaped by the environment it experiences, it follows that what an animal learns depends in part on past niche construction.

A few species, including many vertebrates, have also evolved a capacity to learn from other individuals, and to transmit some of their own learned knowledge to others. We regard this socially learned knowledge as a meme, or meme complex. In humans, this ability to learn from others is facilitated by a further set of processes (such as language and complex cognition), which collectively underlie culture. Within a population, individuals share at least some of their learned knowledge with others, within and between generations. Cultural inheritance probably requires that organisms can decompose their store of cultural 'knowledge' into discrete transmittable 'chunks', perhaps equivalent to psychologist's 'schemata', either in simple or compound form (Holland *et al.* 1986; Plotkin 1996). From our perspective, the term 'meme' is a label for any item of knowledge, or any 'chunk' of such items that is socially learned. As many animals can learn socially, for us, memes are not exclusively human. Obviously, cultural knowledge underlies a great deal of human niche construction. In addition, the environment constructed by humans in part determines which cultural knowledge individuals acquire.

Animal social learning

Modern culture did not suddenly emerge from some precultural Hominid ancestor (Plotkin 1996). The psychological processes and abilities that underlie culture have evolved over millions of years, and can often be found in rudimentary form in animal social learning. Hence, a first step towards an understanding of the evolution of the meme is to consider the nature and evolution of social learning.

Social learning occurs when an animal learns a behaviour pattern or acquires a preference as a consequence of observing, or interacting with, a second animal. The term 'social learning' is a general term that represents learning that is influenced socially. This stands in contrast to asocial learning, in which behaviour acquisition is not influenced by interaction with others. 'Social learning' should not be confused with 'imitation', which loosely describes one psychological process that can result in social learning. 'Imitation' refers to instances where, by observation of another individual performing an act, an animal is able to reproduce the same motor pattern. Local (or stimulus) enhancement refers to a process in which one animal directs another animal's attention to a location (or object) in the environment. If, as a consequence of this tip-off, the observer expresses an equivalent behaviour to the observed, local enhancement can result in a behaviour pattern spreading through a population. Other terms, such as 'social facilitation,' 'observational conditioning', and 'goal emulation', represent other processes that can result in social learning (see Heyes 1994 for a classification).

Imitation is generally regarded as requiring more complex or advanced psychological processing than local enhancement and other processes leading to social learning, although this is unproven. It has been suggested that imitation and teaching are critical to the stable transmission of learned information (Boyd and Richerson 1985), yet this too is unfounded. On the contrary, numerous animal traditions appear to result from psychologically simple mechanisms (Galef 1988; Lefebvre and Palameta 1988). Susan Blackmore (1999) suggests that, of all the processes that can result in social learning, imitation alone can support the transmission of memes, since it alone results in the learning of a behaviour pattern. She argues that other forms of social learning involve learning about the environment, and require the behaviour to be reconstructed by trial and error. In our view this position is misguided (see also Reader and Laland 1999). When imitation results in social learning it is not the motor pattern that is learned, but rather existing topographically defined behavioural elements, alone or in combination, that are associated with the consequences of the behaviour, in a particular context (Heyes 1995). Moreover, studies of imitation in apes and humans have found that the imitated action is rarely perfect the first time, and often builds on previously performed actions (Custance *et al.* 1995). This implies that, even with imitation, some reconstruction of the behaviour pattern is usually required (Sperber 1996 this volume). Thus, there is no reason either to focus predominantly on imitation as the mediator of meme transmission, or to exclude other forms of social learning. All forms of social learning are potentially capable of propagating memes (Reader and Laland 1999).

There are several well-known examples of animal social learning (see Heyes and Galef 1996, for an overview of the field). Perhaps the most celebrated of all cases is the washing of sweet potatoes by Japanese macaques, in which a young female discovered that she could wash the sand grains off her sweet potatoes in water, and this habit spread throughout the troop. In another famous example, Jane Goodall (1964) reported that infant chimps learned the skills necessary for foraging for termites using stalks and twigs by imitating adults.

In fact, most animal social learning is not from parents to offspring, and does not involve cognitively demanding transmission mechanisms. A more typical example is the acquisition of dietary preferences by rats that attend to cues on the breath of conspecifics (Galef 1996). In general, rats prefer to eat foods that other rats have eaten than alternative novel diets, and this simple mechanism probably maintains short term dietary traditions in rat populations. Experiments with Norway rats exploring the social transmission of dietary preferences along chains of animals has established that the diet choices of animals cannot be predicted from animals' consumption of such food items in the absence of conspecifics (Laland and Plotkin 1991, 1993; Galef and Allen 1995). Diet composition may depend on historical factors, and cannot always be predicted from palatability, profitability, or patterns of reinforcement. In other words, which diet choice memes are acquired depends on which memes are already prevalent in the population.

Another informative example is the spread of milk bottle-top opening in British tits (Hinde and Fisher 1951). These birds learned to peck open the foil cap on milk bottles, and to drink the cream, and this behaviour spread throughout Britain and into continental Europe. Hinde and Fisher found that this behaviour probably spreads by local enhancement, where the tits' attention is drawn to the milk bottles by a feeding conspecific, and after this initial tip-off, they subsequently learn *on their own* how to open the tops. However, further analysis by Sherry and Galef (1984) revealed that, in addition to social learning by local enhancement, milk bottle-top opening could be acquired by other means. They found that this behaviour could also spread if the birds were merely exposed to opened milk bottles, even if there were no other birds present to watch performing the opening behaviour. In this example, it is a bottle-opening meme that underlies the birds' niche-constructing behaviour, which is propagated by local enhancement. However, by creating opened milk bottles, this niche construction biases the selective environment of memes in other birds to favour the opening of bottles, and the acquisition of the meme.

The evolution of the meme

How did the process of human cultural evolution evolve from animal social learning? The term 'social learning' as currently applied to animals describes a ragbag of heterogeneous processes with a variety of functions. A more narrow use of the term would restrict it to those processes that might reasonably be regarded as homologous to processes operating in human social learning, and that mediate a general capacity to acquire information from others. Within this narrow category of social learning, humans probably transmit more information vertically (i.e., between generations, from parents to offspring) than any other species (Hewlett and Cavalli-Sforza 1986). For example, Guglielmino *et al.*'s

(1995) study of variation in cultural traits among 277 contemporary African societies found that most traits examined correlated with cultural (linguistic) history rather than with ecological variables. As these societies reside in a range of different habitats, this finding not only suggests a reliance on vertical cultural transmission, it also implies that many of the memes handed down from parent to offspring are of value in a socially constructed world. In contrast, most animal social learning involves the short-term transmission of information about foods and predators among unrelated individuals (Laland *et al.* 1996b). A comparative perspective thus implies that the earliest forms of social transmission in animals were probably horizontal (i.e., within generations), and that the lineage leading to humans was selected (at least initially) for increasing reliance on vertical transmission.

Recent theoretical analyses imply that a shift from transient horizontal traditions towards increased transgenerational cultural transmission should reflect a greater constancy in the environment over time. Over the past twenty years a variety of mathematical analyses have been conducted, exploring the adaptive advantages of social learning, relative to learning asocially, or expressing an unlearned pattern of behaviour that has been adapted over the course of genetic evolution (e.g. Boyd and Richerson 1985; Laland *et al.* 1996b; Feldman *et al.* 1996). These models suggest that when environments change very slowly, adaptive knowledge should be gained at the level of population genetics, while highly variable environments favour reliance on asocial learning. Intermediate rates of environmental change favour social learning, for instance, when changes are not so fast that the transmitter and receiver of the information experience different environments, but not so slow that appropriate genetically transmitted behaviour could evolve instead. Moreover, within this window of intermediate rates of change, it is generally assumed that vertical cultural transmission is an adaptation to slower rates of environmental change than horizontal cultural transmission, because there is an entire generation separating the learning of parent and offspring, during which time the world could change considerably, while peers or siblings can learn from each other virtually simultaneously.

However, the observation that hominid evolution is characterized by a shift towards increased transgenerational cultural transmission is difficult to reconcile with the traditional evolutionary perspective, since there is no evidence to suggest that environments have become more constant over the last few million years. Moreover, even if they had, other protocultural species would be expected to show more vertical transmission too. However, the increasing reliance of hominids on vertical transmission is consistent with a niche construction perspective, since for us, a significant component of the hominid selective environment is assumed to be self-constructed, and therefore partly self-regulated. Hominid niche construction, heavily reliant on memes, could have favoured further vertical transmission, and more memes.

We suggest that our ancestors constructed niches, including sociocultural niches, in which it 'paid' them to transmit more information to their offspring, because the more an organism controls and regulates its environment, and the environment of its offspring, the greater the advantage of transmitting cultural information from one generation to the next. For instance, by tracking or anticipating the movements of migrating or dispersing prey, populations of hominids may have increased the chances that a specific food source was available in their environments, that the same tools used for hunting would always be needed, and that the skin, bones, and other materials from these animals would always be at hand to use in the manufacture of further tools. Such activities create the kind of stable socially constructed environment in which related technologies, such as food preparation or skin processing methods, would be advantageous from one generation to the next, and could be repeatedly socially transmitted from parent to offspring. It is also possible that, once started, transgenerational cultural transmission may become an autocatalytic process, with greater culturally generated environmental regulation leading to increasing homogeneity of environment as experienced by parent and offspring, favouring further transgenerational information transmission. With new cultural traits responding to, or building on, earlier cultural traditions, niche construction sets the scene for an accumulatory culture. This might result in offspring learning higher order 'packages' of cultural traits from their parents, as appears to be the case in pre-industrial societies (Hewlett and Cavalli-Sforza 1986; Guglielmino et al. 1995). Thus, human niche construction, partly dependent on socially transmitted memes, not only partly shapes the selective environment of human genes, but also the selection environment of memes. Human material culture, in the form of tools, artefacts, and homes, may literally be transmitted from one generation to the next, as one aspect of the ecological inheritance of our species.

Consider the astonishing conservatism of both the Oldowan and Acheulean hominid stone tool kits, both of which survived almost unchanged for approximately one million years (Lewin 1998), despite environmental change. Roche *et al.* (1999) recently discovered some evidence for technical diversity in stone tool production in an early (2.34 Myr) site in Kenya, and on the basis of their data, they argued against a hypothesis of technological stasis. However, for us, it makes the degree of stasis that did ensue even more remarkable, since it suggests that cultural selection processes, probably based on the between-generation transmission of memes, must have repeatedly selected against a great deal of spontaneously generated variation in stone tools. It also raises the possibility that evolved psychological mechanisms may have constrained the type of niche-constructing memes that hominids could acquire. Such processes appear to operate in a manner analogous to the elimination of genetic variation by stabilizing natural selection in population genetics.

In post-industrial societies, the acceleratory nature of this accumulatory cultural process may now be causing yet further changes in meme transmission systems in humans, possibly by favouring horizontal cultural transmissions once again. Modern culturally constructed environments appear to be changing so rapidly that, increasingly, vertically transmitted information between parents and offspring is too slow to be of sufficient adaptive value. Nonetheless, the processes remain the same: niche construction, underpinned by a variety of types of information, including memes, modifies the environments that humans experience, which feeds back to shape the type of information, including memes, acquired by individuals and populations.

Although it is certainly not the whole story, the transition from animal protoculture to human culture may perhaps be characterized as two shifts. The first shift is from the horizontal transmission of transient memes adaptive in rapidly changing animal environments, in which the influence of niche construction is only modest, towards transgenerational transmission of stable memes and the development of an accumulatory culture, in environments in which the influence of hominid niche construction is greater. The second is a shift back again to horizontal and oblique transmission in modern times, but now in response to a still-accelerating rate of environmental change caused by the cumulative effects of human meme-based niche construction. Our overall point is that a better understanding of how memes have been transmitted among humans relative to different kinds of selective environments, at different times during our past evolution, could throw some extra light on the evolution of culture itself.

The niche-constructed environment of the meme

What determines whether a meme will spread? For Dawkins (1976), memes, like all replicators, spread if they have fidelity, fecundity, and longevity. In memetic discussions, each of these properties is usually treated as if it is an intrinsic characteristic of the meme. This has resulted in some neglect, even denial, of the capacity of humans to select which memes they adopt, and of the cultural selection processes that themselves determine which particular memes spread (Rose 1998). In spite of an explicit analogy between memes and viruses (Dawkins 1976), memetics as a discipline has tended to concentrate almost exclusively on 'infectiousness' as the factor most responsible for why memes spread. However, the success of a virus depends not only on its *infectiousness*, but also on the *susceptibility* of its hosts, and on whether the *social environment* promotes contact between hosts (Ewald 1994). Based on our evolutionary perspective, we suggest that the same three factors may determine the success of memes.

Our multiple-processes in evolution model explicitly acknowledges that cultural processes build on information acquired through biological evolution and asocial learning. This 'prior' knowledge often shapes the susceptibility of each individual to adopting a particular meme. While the variants that occur during genetic evolution (i.e., mutations), are random (or at least, blind relative to natural selection), those generated and acquired through ontogenetic and cultural processes are 'smart' variants, informed by a priori biases (Seligman 1970; Bolles 1970). Moreover, observations of children (Yando *et al.* 1978) and apes (Russon and Galdikas 1995) suggest that competence guides selection of which actions to imitate. Thus, every individual differs in his or her susceptibility to adopting particular memes depending on genotype, development, individual experience, and social environment, and this susceptibility is not itself exclusively the product of past meme adoption.

In addition to any meme selection on the part of individuals, there is frequently a prior bout of meme selection that occurs in the social domain, as a result of cultural selection processes. There is empirical evidence that the processes of cultural selection sometimes differ from natural selection, and are dependent on aspects of the social environment. For example, studies of social learning in species as diverse as rats, pigeons, and guppies suggest that these animals sometimes adopt a 'do-what-the-majority-do' strategy (Laland *et al.* 1996b). In such cases, the probability that an individual will adopt a meme depends not on its infectiousness, but on the number of individuals already expressing the behaviour. Similar kinds of conformity are prevalent in human societies (Boyd and Richerson 1985). If some strategies are widespread they are likely to generate a conformist transmission, which may act to prevent otherwise more infectious novel memes from invading.

'Do-what-the-successful-individuals-do' is another strategy that individuals in some species adopt, and which imposes biases on meme transmission. For example, bats that are unsuccessful in locating food alone follow previously successful bats to feeding sites (Wilkinson 1992). Starlings can use the foraging success of other birds to assess patch quality, and exploit this information in their judgements as to whether to stay or switch patches (Templeton and Giraldeau 1996). For redwing blackbirds, the social learning of a food preference is affected by whether the demonstrator bird becomes sick or remains well (Mason 1988). Observations of the spread of innovations in primates suggest that whether novel behaviour patterns spread often depends on the identity of the inventor (Kummer and Goodall 1985). In such cases, whether a meme spreads depends on whether successful, charismatic, or powerful individuals adopt it.

However, critics such as Midgley (1994) surely go too far if they deny that the infectiousness of an idea affects its likelihood of being accepted. There is no doubt that memes differ in their attractiveness, their visibility, and their memorability, and *all other things being equal,* the memes with the highest fidelity, fecundity, and longevity will prevail (Dawkins 1976; Blackmore 1999).

Mathematical models for memetics

Few would dispute that evolutionary biology has greatly benefited from the discipline and insights of theoretical population genetics. Any understanding of cultural evolution is likely to benefit in a similar way through the development of a branch of theoretical population memetics. It may surprise some to know that such a body of theory already exists, and has been successfully employed in the study of cultural change and human evolution. Before Dawkins had coined the term 'meme', Cavalli-Sforza and Feldman (1973) were developing population genetics models to explore the processes by which cultural traits spread through populations, and to investigate the coevolution of genes and culture. This work established a small industry of researchers, notably Boyd and Richerson, Aoki, and Rogers, investigating cultural evolution with mathematical models (for a review, see Feldman and Laland 1996). Gene–culture coevolutionary theory is a related branch of theoretical population genetics, which models the interaction between genes and memes throughout the course of human evolution. Whether meme evolution occurs exclusively at the cultural level or through meme–gene interaction, a body of formal theoretical work already exists that can be used to explore memetic processes, test hypotheses, and model data.

The coevolution of lactose absorption and dairy farming represents a good example of meme–gene interaction. Most adult humans are lactose malabsorbers: that is, their level of enzyme (lactase) activity is insufficient to break down the lactose in milk, and its consumption typically leads to sickness. Genetic differences are largely responsible for the difference between absorbers and malabsorbers. A correlation exists between incidence of lactose absorption and history of dairy farming in populations, with absorbers reaching frequencies of over 90% in such populations, but typically less than 20% in populations without dairy traditions. Since milk products have been an important component of the diets of some human populations for over 6000 years, it is conceivable that agricultural niche construction in the form of dairy farming may have created the selective regime under which the genes for absorption were favoured.

Feldman and Cavalli-Sforza (1989) used gene–culture coevolutionary theory to investigate the evolution of lactose absorption. By defining genotypes that differ in terms of their ability to process lactose, and by describing individuals as either having a meme for milk consumption or not, Feldman and Cavalli-Sforza were able to develop a population genetics model to explore how dairy farming and milk use might coevolve with genes for lactose absorption. The analysis suggested that whether or not the allele for absorption achieves a high frequency depends critically on the probability that the children of milk users themselves adopt the meme. The analysis is able to account for both the spread of lactose absorption, and the culturally related variability in its incidence. Moreover, Feldman and Cavalli-Sforza found a broad range of conditions under which the absorption allele does not spread despite a significant fitness advantage. Meme transmission complicates the selection process to the extent that the outcome may differ from that expected under purely genetic transmission.

This work, and numerous other studies, simply would not have been possible without the assumption that culture could be broken down into discrete units, akin to memes. There already exists a respectable, and well-established, formal theory of memetics, in the form of cultural evolutionary and gene–culture coevolutionary theory (Cavalli-Sforza and Feldman 1981; Boyd and Richerson 1985; Feldman and Laland 1996). We recommend that meme enthusiasts exploit it.

Conclusion

A focus on niche construction aids an understanding of how the ideational, behavioural and material components of culture could evolve. Organisms, though their niche construction, play an important role in the evolutionary process by modifying the selection pressures acting on their genes. In the human case, niche construction, informed by a variety of information-gaining processes, modifies the environment in which both memes and genes are selected. Human material culture may be regarded as one aspect of the ecological inheritance of our particular species. Among the most successful of memes are those expressed in niche construction, which effectively bias their selective environment in their own favour.

Acknowledgements

Kevin Laland is supported by a Royal Society University Research Fellowship. We are grateful to Gillian Brown for helpful comments on an earlier draft of this manuscript.

References

Blackmore, S. (1999). *The meme machine.* Oxford: Oxford University Press.

Bolles, R. C. (1970). Species-specific defence reactions and avoidance learning. *Psychological Review,* 77: 32–48.

Boyd, R. and Richerson, P.J. (1985). *Culture and the evolutionary process,* Chicago: University of Chicago Press.

Cavalli-Sforza, L. L. and Feldman, M. W. (1973). Cultural versus biological inheritance: Phenotypic transmission from parent to children (a theory of the effect of parental phenotypes on children's phenotype). *American Journal of Human Genetics,* 25: 618–37.

Cavalli-Sforza, L. L. and Feldman, M. W. (1981). *Cultural transmission and evolution: A quantitative approach,* Princeton: Princeton University Press.

Custance, D. M., Whiten, A., and Bard, K. A. (1995). Can young chimpanzees (Pan troglodytes) imitate arbitrary actions? Hayes and Hayes (1952) revisited. *Behaviour,* 132: 837–59.

Darwin, C. (1881). *The formation of vegetable mold through the action of worms, with observations on their habits.* London: John Murray.

Dawkins, R. (1976). *The selfish gene.* Oxford: Oxford University Press.

Dawkins, R. (1982). *The extended phenotype.* Oxford: Oxford University Press.

Dennett, D. C. (1995). *Darwin's dangerous idea.* London: Penguin.

Ewald, P. W. (1994). *Evolution of infectious diseases.* Oxford: Oxford University Press.

Feldman, M. W., Aoki, K. and Kumm, J. (1996). Individual versus social learning: Evolutionary analysis in a fluctuating environment. *Anthropological Science,* 104(3): 209–32.

Feldman, M. W. and Cavalli-Sforza, L. L. (1976). Cultural and biological evolutionary processes, selection for a trait under complex transmission. *Theoretical Population Biology,* 9(2): 238–59.

Feldman, M. W. and Cavalli-Sforza, L. L. (1989). On the theory of evolution under genetic and cultural transmission with application to the lactose absorption problem, In *Mathematical Evolutionary Theory* (ed. M. W. Feldman). Princeton: Princeton University Press, pp. 145–73.

Feldman, M. W. and Laland, K. N. (1996). Gene-culture coevolutionary theory. *Trends in Ecology and Evolution,* 11: 453–7.

Forshaw, J. (1998). *Encyclopedia of Birds* (2nd edn) San Diego: Academic Press.

Galef, B.G. Jr. (1988). Imitation in animals: History, definition, and interpretation of data from the psychological laboratory. In *Social learning: Psychological and biological perspectives* (ed. T. R. Zentall, and B. G. Galef Jr.). Hillsdale, NJ: Erlbaum, pp. 3–28.

Galef, B. G. Jr. (1996). Social enhancement of food preferences in Norway rats: A brief review. In *Social Learning in Animals: the Roots of Culture* (ed. Heyes, C. M. and Galef, B. G. Jr.), pp 49–64. San Diego: Academic Press.

Galef, B. G. Jr. and Allen, C. (1995). A new model system for studying behavioural traditions in animals. *Animal Behaviour,* 50(3): 705–17.

Goodall, J. (1964). Tool using and aimed throwing in a community of free living chimpanzees. *Nature,* 201: 1264–6.

Guglielmino, C. R., Viganotti, C., Hewlett, B., and Cavall-Sforza, L. L. (1995). Cultural variation in Africa: Role of mechanisms of transmission and adaptation. *Proceedings of the National Academy of Science USA*, **92**: 7585–9.

Gullan, P. J. and Cranston, P. S. (1994). *The insects. An outline of entomology*. London: Chapman & Hall.

Hansell, M. H. (1984). *Animal architecture and building behaviour*. New York: Longman.

Hewlett, B. S. and Cavalli-Sforza, L. L. (1986). Cultural transmission among Aka pygmies. *American Anthropologist*, **88**: 922–34.

Heyes, C. M. (1994). Social learning in animals: categories and mechanisms. *Biological Reviews*, **69**: 207–31.

Heyes, C. M. (1995). Imitation and flattery: A reply to Byrne and Tomasello. *Animal Behaviour*, **50**, 1421–24.

Heyes C. M. and Galef, B. G. Jr. (ed.) (1996). *Social learning in animals: The roots of culture*. San Diego: Academic Press.

Hinde, R. A. and Fisher, J. (1951). Further observations on the opening of milk bottles by birds. *British Birds*, **44**: 393–6.

Holland, J. H., Holyoak, K. J., Nisbett, R. E. and Thagard, P. R. (1986). *Induction. Processes of inference learning and discovery*. Cambridge, MA: MIT Press.

Holldobler, B. and Wilson, E. O. (1994). *Journey to the ants. A story of scientific exploration*. Cambridge, MA: Belknap.

Jones, C. G., Lawton, J. H. and M. Shachak (1997). Positive and negative effects of organisms as physical ecosystem engineers. *Ecology*, **78**: 1946–57.

Kirkpatrick, M. and Lande, R. (1989). The evolution of maternal characters. *Evolution*, **43**(3): 485–503.

Kummer, H. and Goodall, J. (1985). Conditions of innovative behaviour in primates. *Philosophical Transactions of the Royal Society of London Series B*, **308**: 203–14.

Laland, K. N. and Plotkin, H. C. (1991). Excretory deposits surrounding food sites facilitate social learning of food preferences in Norway rats. *Animal Behaviour*, **41**: 997–1005.

Laland, K. N. and Plotkin, H. C. (1993). Social transmission in Norway rats via excretory marking of food sites. *Animal Learning and Behavior*, **21**: 35–41.

Laland, K. N., Odling-Smee, F. J. and Feldman, M. W. (1996a). On the evolutionary consequences of niche construction. *Journal of Evolutionary Biology*, **9**: 293–316.

Laland, K. N., Richerson, P. J. and Boyd, R. (1996b). Developing a theory of animal social learning, In *Social learning in animals: The roots of culture* (ed. C. M. Heyes and B. G. Galef Jr.), pp. 129–54. San Diego: Academic Press.

Laland, K. N., Odling-Smee F. J. and Feldman M. W. (1999). The evolutionary consequences of niche construction and their implications for ecology. *Proceedings of the National Academy of Science USA*, **96**(18): 10242–7.

Laland, K.N., Odling-Smee, F.J. and Feldman, M. W. (2000). Niche construction, biological evolution and cultural change. *Behavioral and Brain Sciences*, **23**(1): 131–75.

Lee, K. E. (1985). *Earthworms: their ecology and relation with soil and land use*. London: Academic Press.

Lefebvre, L., and Palameta, B. (1988). Mechanisms, ecology and population diffusion of socially learned food finding behavior in feral pigeons. In *Social learning: Psychological and biological perspectives* (ed. T. Zentall, and B.G. Galef Jr.), pp. 141–164. Hillsdale, NJ: Erlbaum.

Lewin, R. (1998). *Principles of human evolution.* Malden, MA: Blackwell.

Lewontin, R. C. (1983). Gene, organism, and environment. In *Evolution from Molecules to Men* (ed. D. S. Bendall) Cambridge: Cambridge University Press, pp 273–83.

Lewontin, R. C. (2000). *The Triple Helix.* Cambridge MA: Harvard University Press.

Mason, J. R. (1988). Direct and observational learning by redwing blackbirds (*Agelaius phoeniceus*): The importance of complex visual stimuli. In *Social learning: Psychological and biological perspectives* (ed. T. Zentall, and B. G. Galef Jr.), pp. 99–115. Hillsdale, NJ: Erlbaum.

Midgley, M. (1994). Letter to the editor. *New Scientist,* 12 February, 50.

Nowak, R. M. (1991). *Walker's mammals of the world* (5th edn). Baltimore: The Johns Hopkins University Press.

Odling-Smee, F. J. (1988). Niche constructing phenotypes. In *The role of behavior in evolution* (ed. H. C. Plotkin) Cambridge, MA: MIT Press.

Odling-Smee, F. J., Laland, K. N. and Feldman, M. W. (1996). Niche construction. *The American Naturalist,* 147(4): 641–48.

Paxton J. R., and Eschmeyer W. N. (1998). *Encyclopedia of fishes.* San Diego: Academic Press.

Plotkin, H. C. (1996). Non-genetic transmission of information: Candidate cognitive processes and the evolution of culture. *Behavioral Processes,* 35: 207–13.

Plotkin, H. C. and Odling-Smee, F. J. (1981). A multiple-level model of evolution and its implications for sociobiology. *Behavioral and Brain Sciences,* 4: 225–68.

Preston-Mafham, R. and Preston-Mafham, K. (1996). *The natural history of insects.* Swindon, UK: Crowood Press.

Reader, S. M. and Laland, K. N. (1999). Do animals have memes? *Journal of Memetics– Evolutionary Models of Information Transmission,* 3. [http://www.cpm.mmu.ac.uk/ jom-emit/1999/vol3/reader_sm&laland_kn.html]

Robertson, D. S. (1991). Feedback theory and Darwinian evolution. *Journal of Theoretical Biology,* 152: 469–84.

Roche, H. *et al.* (1999). Early hominid stone tool production and technical skill 2.34 Myr ago in West Turkana, Kenya. *Nature,* 399: 57–60.

Rose, N. (1998). Controversies in meme theory. *Journal of memetics–Evolutionary Models of Information Transmission,* 2. [http://www.cpm.mmu.ac.uk/jom-emit/1998/vol2/ rose_n.html]

Russon, A. E. and Galdikas, B. M. F. (1995). Constraints on great apes imitation: Model and action selectivity in rehabilitant organgutan (*Pongo pygmaeus*) imitation. *Journal of Comparative Psychology,* 109(1), 5–17.

Seligman, M. E. P. (1970). On the generality of the laws of learning. *Psychological Review,* 77: 406–18.

Sherry, D. F. and Galef, B. G., Jr. (1984). Cultural transmission without imitation— milk bottle opening by birds. *Animal Behaviour,* 32: 937–8.

Sperber, D. (1996). *Explaining culture: A naturalistic approach.* Oxford: Blackwell.

Szathmáry, E. and Maynard Smith, J. (1995). The major evolutionary transitions. *Nature,* 374: 227–31.

Templeton J. J. and Giraldeau, L. A. (1996). Vicarious sampling: The use of personal and public information by starlings foraging in a simple patchy environment. *Behavioural, Ecology and Sociobiology,* 38: 105–13.

Tylor, E. B. (1871). *Primitive culture.* London.

Wilkinson, G. (1992). Information transfer at evening bat colonies. *Animal Behaviour,* **44**: 501–18.

Yando, R., Seitz, V. and Zigler, E. (1978). *Imitation: A developmental perspective.* Hillsdale, NJ: Erlbaum.

CHAPTER 7

Memes: Universal acid or a better mousetrap?

Robert Boyd and Peter J. Richerson

Among the many vivid metaphors in *Darwin's Dangerous Idea*, one stands out. The understanding of how cumulative natural selection gives rise to adaptations is, Dennett says, like a 'universal acid'—an idea so powerful and corrosive of conventional wisdom that it dissolves all attempts to contain it within biology. Like most good ideas, this one is very simple: once replicators (material objects that are faithfully copied) come to exist, some will replicate more rapidly than others, leading to adaptation by natural selection. The great power of the idea is that the resulting adaptations can be understood by asking what leads to efficient, rapid replication. Given that ideas seem to replicate, it is natural that Dawkins (1976, 1982), Dennett (1995), and others have explored the possibility of using this idea to explain cultural evolution.

Natural selection was not Darwin's only powerful, far-reaching idea. Ernst Mayr (1982) has argued that what he calls 'population thinking' was also among Darwin's foundational contributions to biology. Before Darwin, species were thought to be essential, unchanging types, like geometric figures and chemical elements. Darwin saw that species were populations of organisms that carried a variable pool of inherited information through time. To understand the evolution of species, biologists had to account for the processes that changed the nature of that inherited information. Darwin thought that the most important processes were natural selection, sexual selection, and the 'inherited effects of use and disuse.' We now know that the last process is not important in organic evolution—unlike Darwin, modern biologists do not believe that the sons of blacksmiths inherit their father's mighty biceps. Nowadays biologists think many processes that Darwin never dreamed of are important including segregation, recombination, gene conversion, and meiotic drive. Nonetheless, modern biology is fundamentally Darwinian because its explanations of evolution are rooted in population thinking. If Darwin were to be resurrected tomorrow through some miracle of cloning, we think he would be quite happy with his legacy.

In this chapter we want to convince you that population thinking, not natural selection, is the key to conceptualizing culture in terms of material causes. This argument is based on three well-established facts:

1. *There is persistent cultural variation among human groups.* Any explanation of human behavior must account for how this variation arises and how it is maintained.

2. *Culture is information stored in human brains.* Every human culture contains vast amounts of information. Important components of this information are stored in human brains.

3. *Culture is derived.* The psychological mechanisms that allow culture to be transmitted arose in the course of hominid evolution. Culture is not simply a by-product of intelligence and social life.

Much of culture is information stored in human brains—information that got into those brains by various mechanisms of social learning. It follows that to explain the distribution of information stored in the brains of the members of current generation, any coherent theory will have to account for the cultural information in the brains of the previous generation. The theory will also have to explain how this information, together with genes and environmental contingencies, caused the present generation to acquire the cultural information that it did. Unfortunately, we do not understand how this process works. It may be that cultural information stored in brains takes the form of discrete memes that are replicated faithfully in each subsequent generation, or it may not. This is an empirical question that at present is unanswered, and we will see that other models are possible. In every case, the Darwinian population approach will illuminate the process by which the cultural information that is stored in a population of brains is transformed from one generation to the next.

We also want to convince you that population thinking can play an important, constructive role in the human sciences. The fact that population thinking is logically necessary for a natural, causal, theory of culture, does not necessarily mean that such a theory will be useful. Thus, we know that human culture must be consistent with quantum mechanics, but it is unlikely that such a connection will help us understand, say, ethnic conflict. However, we think Darwinian models of culture *are* useful for two reasons. First, they serve to connect the rich models of behavior based on individual action developed in economics, psychology, and evolutionary biology with the data and insights of the cultural sciences, anthropology, archaeology, and sociology. In doing so, we think that they can help shed light on important unsolved problems in the social sciences. Second, population thinking is useful because it offers a way to build a mathematical theory of human behavior that captures the important role of culture in human affairs. Population thinking is not a universal acid that will dissolve existing social sciences. But, it is a better mousetrap, providing useful new tools that can help solve outstanding problems in the human sciences.

Culture is heritable at the group level

One of the striking facts about the human species is that there are important, persistent differences between human groups that are created by culturally transmitted ideas, not genetic differences, or differences in the physical or biotic environment. Sonya Salamon's (1992) research on immigrant communities in the United States shows how cultural differences can give rise to different behaviors in the same environment. One of Salamon's studies focused on two farming communities in southern Illinois. 'Freiburg' (a pseudonym), is inhabited by the descendents of German-Catholic immigrants who arrived in the area during the 1840s. 'Libertyville' (also a pseudonym) was settled by people from other parts of the United States—mainly Kentucky, Ohio, and Indiana—when the railroad arrived in 1870. These two communities are only about twenty miles apart and have been carefully matched for similar soil types.

The people in these two communities have different values about family, property, and farm practice, and these differences seem consistent with their ethnic origins. The farmers of Freiburg tend to value farming as a way of life, and they want at least one son or daughter to continue as a farmer. In Freiburg, wills specify that the farm will go to a child who will farm the land and use farm proceeds to buy out any non-farming siblings. Parents put considerable pressure on children to become farmers. They place little importance on education, knowing that advanced education often results in young people not returning to farm. Salomon argues that these 'yeoman' values are similar to those observed among peasant farmers in Europe and elsewhere. In contrast, the 'Yankee' farmers of Libertyville regard their farms as profit-making businesses. They buy or rent land depending on economic conditions and if the price is right they sell. Many Yankee farmers would prefer their children to continue farming, but they see it as an individual decision. Some families help their children enter farming, but many do not, and they generally place a strong value on higher education.

The difference in values between Freiburg and Libertyville lead to measurable differences in farm practices despite the proximity of the two towns and the similarity of their soils. Farms are substantially larger in Libertyville—the mean size of farm operations in Libertyville is 518 acres compared to 276 acres in Freiburg. The Libertyville farms are larger because Yankee farmers rent more land. They rent more land because Yankees demand a higher income to stay in farming. Yeomen, who so value farming for its own sake, are content with lower incomes and fear the risks of debt-financed expansion.

The two communities also show striking differences in farm operations. In Libertyville, as in most of southern Illinois, farmers specialize in grain production. It is the primary source of income for 77% of the farmers in Libertyville. In Freiburg, many people mix grain production with dairying or livestock-raising, activities that are almost absent in Libertyville. Because animal husbandry is labor-intensive, it allows Germans to accommodate their larger families on their more limited acreage. Yankee farmers decided against dairying and stock raising because grain farming is more profitable and less work.

The fact that culturally distinctive human groups behave differently in the same environment implies that culture is heritable, at least at the group level. Many beliefs and values that are common in a group at one point in time are also common among the descendants of the same group. Any theory of how culture works must be consistent with this fact. It must explain why the German farmers of Freiburg hold different beliefs about life and land than their Yankee neighbors almost 150 years after leaving Europe.

Culture is information in stored human brains

Every human culture contains an enormous amount of information. Consider how much information must be transmitted to maintain a particular distinctive spoken language. A lexicon requires something like 10 000 associations between words and their meanings. Grammar entails a complex set of rules regulating morphosyntax, and although it is unclear the extent to which these rules arise from innate, genetically transmitted structures, it is clear that the rules that underlie the grammatical differences that separate English and Chinese are culturally transmitted. Subsistence techniques also entail large amounts of information. For example, Blurton-Jones and Konner (1976) showed that the !Kung San have a very detailed knowledge of the natural history of the Kalahari—so detailed, in fact, that the researchers were unable to judge the accuracy of much of !Kung knowledge because in some aspects it exceeded Western biology. As anyone who as ever tried to make a decent stone tool can attest, the manufacture of even the simplest tool requires lots of knowledge; more complex technologies require even more. Imagine the instruction manual for constructing a seaworthy kayak from materials available on the North Slope of Alaska. The institutions that regulate social interactions incorporate still more information. Property rights, religious custom, roles, and obligations all require a considerable amount of detailed information.

The vast store of information that exists in every culture cannot simply float in the air. It must be encoded in some material object. In societies without widespread literacy, the most important objects in the environment capable of storing this information are human brains and human genes. It is undoubtedly true that some cultural information is stored in artifacts. It may well be that the designs that are used to decorate pots are stored on the pots themselves, and that when young potters learn how to make pots they use old pots, not old potters, as models. In the same way, the architecture of the church may help store information about the rituals performed within. Without writing, however, the ability of artifacts to store culture is quite limited. First, many artifacts are very difficult to reverse-engineer. The young potter cannot learn how to select clay and temper, or how to fire a pot by studying existing ones. Second, much cultural information is semantic knowledge—how can an artifact store the notion that Kalahari porcupines are monogamous? Or the rules that govern bride-price transactions?

It is also clear that much cultural information is not stored in human genes. In one sense this is obvious. The evidence is very clear that very little cultural variation results from genetic differences. We know that genetic differences do not explain why some people speak Chinese and others English, or why the !Kung know a lot more about the biology of porcupines than most readers of this chapter.

However, there is a subtle and much more plausible way that genes could store cultural information. It could be that most human culture is innate, genetically transmitted information that is evoked by environmental cues. Pascal Boyer (1994) argues that much of religious belief has this character. For example, the Fang, a group Boyer studied in Cameroon, have elaborate beliefs about ghosts. For the Fang, ghosts are malevolent beings that want to harm the living; they are invisible and can pass through solid objects, and so on. Boyer argues that most of what the Fang believe about ghosts is not culturally transmitted; rather it is based on the innate, epistemological assumptions that underlie all cognition. Once a young Fang child learns that ghosts are sentient beings, she does not need to learn that ghosts can see or that they have beliefs and desires—these components are provided by cognitive machinery that reliably develops in every environment. According to this view, cultural differences arise because different environmental cues evoke different innate information. A friend of ours believes in angels instead of ghosts because he grew up in an environment in which people talked about angels. However, most of what he knows about angels comes from the same cognitive machinery that gives rise to Fang beliefs about ghosts, and the information that controls the development of this machinery is stored in the genome.

This picture of culture is a useful antidote to the simplistic view that culture is simply poured from one head into another. Evolutionary psychologists are surely right that every form of learning, including social learning, requires an information-rich innate psychology, and that much of the adaptive complexity we see in cultures around the world stems from this information. However, it is a big mistake to ignore transmitted cultural information. The single most important adaptive feature of culture is that it allows the gradual, cumulative assembly of adaptations over many generations—adaptations that no single individual could invent on their own. Cumulative adaptation cannot be based solely on innate, genetically encoded information.

Consider the evolution of a relatively simple form of technology, the mariners' magnetic compass (Needham 1978). First, Chinese geomancers noticed the peculiar tendency of small magnetite objects to orient in the earth's magnetic field, an effect that they used for purposes of divination. Then, Chinese mariners learned that magnetized needles could be floated on water to indicate direction at sea. Next, over several centuries Chinese seamen developed a dry compass mounted on a vertical pin-bearing, like a modern toy compass. Europeans acquired this type of compass in the late medieval period. European seamen then developed the fixed card compass that allowed a helmsman to steer an accurate course by aligning the bow mark with the appropriate compass point. Compass makers later learned to adjust iron balls near the compass to zero out the magnetic influence from the ship, and to gimbal the compass and fill it with liquid to damp the motion imparted to the card by the roll and pitch of the ship. Even such a relatively simple tool was the product of at least seven or eight innovations separated in time by centuries and in space by the breadth of Eurasia. This sort of adaptation only occurs because novel information can accumulate in human populations, be stored in human brains, and be transmitted through time by teaching and imitation.

Evolutionary psychologists argue that our psychology is built of complex, information-rich, evolved modules that are adapted for the hunting and gathering life that we pursued until the origins of agriculture a few thousand years ago. On this argument, humans can easily and naturally do the things we are really adapted to do like learn a language or understand the feelings of others. Inventing complex modern artefacts like the compass is hard, but what about skills necessary for hunting and gathering? Couldn't we learn these as easily as we learn language? Doesn't our brain contain the information necessary to follow hunting and gathering ways? Our ancestors lived as hunter-gatherers of some kind for the last 2 or 3 million years. If we had to do so, couldn't we reinvent that stuff, just as Fang children invent the properties of their ghosts, or children can invent a grammar?

Good questions, but we think the answer is almost certainly 'Are you nuts?!' Consider the following thought experiment. Suppose you are stranded in some not-too-extreme desert environment, not the Empty Quarter or the Atacama, but the desert between Sonoita, Mexico and Yuma, Arizona. Your task is to survive and raise your kids without modern technology. You will be given the resources to survive a few months to get your feet on the ground before we take away your last tin of food and your last steel tool—a little time to see what comes naturally. Will you make it?

We don't think so. The stretch between Sonoita and Yuma is known as El Camino del Diabolo, 'the Devil'sRoad.' It was one leg of the main overland route from Old Mexico to California until the coming of railroads. For more than a century it was used by Spanish, Mexican, and American travelers. To get that far, every traveler had to already be an experienced frontiers-person, and no doubt most were hardbitten, desert-wise, and well equipped with familiar technology. It was the best of several bad routes and was comparatively well known and well marked. Still, it was an infamous leg of the journey, and many travelers ended up in the hasty graves that litter the route.

Now, consider that the Camino del Diabolo was also the home to Papago indians who, with a few pounds of wood, stone and bone equipment, an impressive amount of hard-won knowledge, and a well-adapted system of social institutions, lived and raised their children in very same desert that killed so many pioneers. If our task was to survive in this desert without our accustomed industrial technology, we would certainly trade a few hours of tutoring by a traditional Papago for any number of months trying to summon an innate knowledge of the desert.

Culture is derived

Simple forms of social learning, often termed 'protoculture,' occur in many other species of animals. In a review of the social transmission of foraging behavior, Levebre and Palameta (1988) give 97 examples of protocultural variation in foraging behavior in animals as diverse as baboons, sparrows, lizards, and fish. Much of the evidence for protoculture in other animals consists of observations of different behavior by populations of the same species living in similar environments. For example, chimpanzees in the Mahale Mountains of Tanzania often adopt a unique grooming posture in which both partners extend one arm over their heads, clasp hands, and then groom one another's exposed arm pits. These grooming hand-clasps occur often and are performed by all members of the group. Chimpanzees at Gombe, who live less than 100 kilometers away in a similar type of habitat, often groom but never perform this behavior. Sometimes scientists have observed the spread of a novel behavior. One famous example comes from Japan where a group of Japanese macaques, whose range included a sandy beach, were provisioned with sweet potatoes. A young female macaque accidentally dropped her sweet potato into the sea as she was trying to rub the sand off it. She must have liked the result, as she began to carry all of her potatoes to the sea to wash them. Other monkeys followed suit. However, it took other members of the group quite some time to acquire the behavior and many monkeys never washed their potatoes. Finally, some evidence for protoculture in other animals comes from experiments which demonstrate that behavior is socially transmitted. The most famous case is the transmission of song dialects in birds like the white-crowned sparrow.

There is little evidence, however, of cumulatively evolved cultural traditions in other species. With a few exceptions, social learning leads to the spread of behaviors that individuals could have learned on their own. For example, food preferences are socially transmitted in rats. Young rats acquire a preference for a food when they smell the food on the pelage of other rats (Galef 1988). This process can cause the preference for a new food to spread within a population. It can also lead to behavioral differences among populations living in the same environment, because current foraging behavior depends on a history of social learning. However, it does not lead to the cumulative evolution of complex new behaviors that no individual rat could learn on its own. Thus, in other animals it is quite plausible that most of the detailed information that creates protocultural differences is stored and transmitted genetically.

Circumstantial evidence suggests that the ability to acquire novel behaviors by observation is essential for cumulative cultural change. Students of animal social learning distinguish *observational learning* which occurs when younger animals observe the behavior of older animals and learn how to perform a novel behavior by watching them, from a number of other mechanisms of social transmission which also lead to behavioral continuity without observational learning (Galef 1988; Visalberghi and Fragazy 1990; Whiten and Ham 1992). One such mechanism, *local enhancement*, occurs when the activity of older animals increases the chance that younger animals will learn the behavior on their own. Imagine a young monkey acquiring its food preferences as it follows its mother around. Even if the young monkey never pays any attention to what its mother eats, she will lead it to locations where some foods are common and others rare, and the young monkey may learn to eat much the same foods as mom.

Local enhancement and observational learning are similar in that they can both lead to persistent behavioral differences among populations, but only observational learning allows *cumulative* cultural change (Tomasello *et al.* 1993). To see why, consider the cultural transmission of stone tool use. Suppose that occasionally early hominids learned to strike rocks together to make useful flakes. Their companions, who spent time near them, would be exposed to the same kinds of conditions and some of them might learn to make flakes too, entirely on their own. This behavior could be preserved by local enhancement because groups in which tools were used would spend more time in proximity to the appropriate raw materials. However, that would be as far as tool-making would go. Even if an especially talented individual found a way to improve the flakes, this innovation would not spread to other members of the group because each individual learned the behavior anew, without any detailed guidance from innovators who have improved on the common technique. Local enhancement is limited by the learning capabilities of individuals and the fact that each new learner must start from scratch. With observational learning, on the other hand, innovations can be incorporated into others' behavioral repertoires if younger individuals are able to acquire the improved behavior by observational learning. To the extent that observers can use the behavior of models as a starting point, observational learning can lead to the cumulative evolution of behaviors that no single individual could invent on its own.

Adaptation by cumulative cultural evolution is apparently not a by-product of intelligence and social life. Capuchin monkeys are among the world's cleverest creatures. They resemble apes in having quite large brains for their size. In nature, they perform many complex behaviors, and in captivity they can be taught extremely demanding tasks. Capuchins live in social groups and have ample opportunity to observe the behavior of other individuals of their own species. Yet good laboratory evidence indicates that these monkeys make little or no use of observational learning (Visalberghi and Fragazy 1990). Observational learning is not simply a by-product of intelligence and the opportunity to observe conspecifics. Rather, it seems to require special psychological mechanisms (Bandura 1986). This conclusion suggests that the psychological mechanisms that enable humans to learn by observation are adaptations which have been shaped by natural selection in the human lineage because culture is beneficial.

Cultural evolution is Darwinian

Now, let us consider what these facts imply for a theory of culture. Consider a population of individuals who are culturally interconnected; they speak dialects of a single language, use similar technology, share relatively similar beliefs about the world, and have similar moral values. People in this population think and behave differently from other peoples, in part, because they have different culturally transmitted information stored in their brains. Next consider the descendants of this population, say 100 years later. The culture of the descendant population will be similar in many ways to that of their predecessors. Their language will be similar, and they may often use similar technology, have similar beliefs about the world and subscribe to a similar moral system. The fact that culture depends on behavior stored in the brains of this population requires us to account for how the information that generates these similarities was transmitted from the brains in the first population to the brains in the second.

Of course, there will also be differences between the two populations, some small, some great. Some of these differences will arise because some behaviors are more common in the second population—for example, perhaps what was previously a rare usage or form of pronunciation has become common. Other differences will arise because genuinely new behavior is present, either as a result of borrowing from neighboring populations or due to genuine innovation. Thus, a complete theory would also have to account for why some forms of cultural information spread, and why some forms have diminished, and how innovation occurs.

Cumulative cultural change requires observational learning. People observe the behavior of others, and (somehow) acquire the information necessary to produce a reasonable facsimile of the same behavior. In any given time period, each person observes only a sample of the people who make up his population. A very small child is exposed mainly to the people in her family, older children are exposed to peers and teachers, and adults to yet a wider range of people. We will refer to this group of people as an individual's 'cultural sample'. For most of human history cultural samples were small, but nowadays they may be immense. On the other hand, for some elements of culture many people may be disproportionately influenced by a single charismatic leader or acknowledged expert.

The fact that cultures often persist over time with little change means that the commonness of a behavior in an individual's cultural sample must have a positive effect on the probability that the individual ultimately acquires the cultural information that generates that behavior. Such a tendency could arise in several different ways: if observational learning takes the form of approximately unbiased copying, then common behaviors will be more frequent in cultural samples, and therefore will be more likely to be copied. It could also be that the psychology of observational learning itself predisposes people to acquire more common behaviors. Finally, it could be that rare behaviors are typically disadvantageous and less likely to be retained as a result of individual learning and experimentation, or even by natural selection against them.

It follows that cultural change is a population process. The argument proceeds in several steps:

- To understand how a person behaves, we have to know the nature of the information stored in her brain.
- To understand why people have the beliefs that they do, we must know what kinds of behaviors characterized their cultural sample.
- To predict the distribution of cultural samples that exists, you must know the cultural composition of the population.
- Therefore, to understand how people behave, we must understand why the population has the cultural composition that it does.

Similarities between descendant and ancestral populations arise because the necessary information has been transmitted from individual to individual through time without significant change. Differences occur because some variants have become more common, others have become more rare, and some completely new variants have been introduced. Thus, to account for both continuity and change we need to understand the population processes by which ideas are transmitted through time.

Culturally transmitted skills and beliefs may not be replicators

In *The Extended Phenotype*, Richard Dawkins (1982) argues that the cumulative evolution of complex adaptations requires what he calls replicators, things in the physical world that produce copies of themselves, and have the following three additional properties:

1. *Fidelity*. The copying must be sufficiently accurate that even after a long chain of copies the replicator remains almost unchanged.
2. *Fecundity*. At least some varieties of the replicator must be capable of generating more than one copy of themselves.
3. *Longevity*. Replicators must survive long enough to affect their own rate of replication.

Replicators give rise to cumulative adaptive evolution because replicators are targets of natural selection. Genes are replicators—they are copied with astounding accuracy, they can spread rapidly, and they persist throughout the lifetime of an organism, directing its machinery of life. Dawkins thinks that beliefs and ideas are also replicators. On the face of it, this is an apt analogy. Beliefs and ideas can be copied from one mind to another, spreading through a population, controlling the behavior of people who hold them.

But there are reasons to doubt that beliefs and skills are replicators, at least in the same sense that genes are. Unlike genes, ideas are not copied and transmitted intact from one brain to another. Instead, the information in one brain generates some behavior, somebody else observes this behavior, and then (somehow) creates the information necessary to generate very similar behavior. The problem is that there is no guarantee that the information in the second brain is the same as the first. For any phenotypic performance there are potentially an infinite number of rules that would generate that performance. Information will be transmitted from brain to brain only if most people induce a unique rule from a given phenotypic performance. While this may often be the case, it is also plausible that genetic, cultural, or developmental differences among people may cause them to infer different beliefs from the same overt behavior. To the extent that these differences shape future cultural change, the replicator model captures only part of cultural evolution.

The generativist model of phonological change illustrates the problem. According to the generativist school of linguistics, individual pronunciation is governed by a complex set of rules that takes as input the desired sequence of words and produces as output the sequence of sounds that will be produced (Bynon 1977). Generativists also believe that, as adults, people can modify their pronunciation only by adding new rules that act at the *end* of the chain of existing rules. Children, on the other hand, are not constrained by the rules used to generate adult speech. Instead, they induce the simplest set of grammatical rules that will account for the performances they hear, and these may be quite different than the rules used by adult speakers. Although the new rules produce the same performance, they can have a different structure, and therefore, allow further changes by rule addition that would not have been possible under the old rules.

The following example (from Bynon 1977) illustrates this phenomenon. In some dialects of English, people pronounce words that begin with *wh* using what linguists call an 'unvoiced' sound while they pronounce words beginning with *w* using a voiced sound. (Unvoiced sounds are produced with the glottis open, resulting in a breathy sound, whereas voiced sounds are produced with the glottis closed, causing a resonant tone.) People who speak such dialects must have mental representations of the two sounds and rules to assign them to appropriate words. Now suppose that people who speak such a dialect come into contact with other people who only use the voiced *w* sound. Further suppose that this second group of people is more prestigious, and accordingly people in the first group modify their speech so that they too use only voiced *w*s. According to the generativists, they will accomplish this change by adding a new rule which says 'voice all unvoiced *w*s.' So, Larry wants to say *Whether it is better to endure . . .* The part of his brain that takes care of such things looks up the mental representations for each of the words including *whether* which has an unvoiced *w* (because that is the way Larry learned to speak as a child). Then after any other processing for stress or tone, the new rule changes the unvoiced *w* in *whether* to a voiced *w*. Children learning language in the next generation never hear an unvoiced *w* and, according to generativists, they adopt the same underlying representation for *whether* and *weather*. Thus, even though there is no difference in the phenotypic performance among parents and children, children do not acquire the same mental representation as their parents. This difference may be important because it will affect further changes. For example, it might make it less likely that the two sounds would split again in the future. The adult version of the rule still has a latent distinction between the voiced and unvoiced pronunciation that could serve as the basis for renewing the distinction whereas, if the generativists are correct, the latent distinction is unavailable to child learners who hear only one usage.

Replicators are not necessary for cumulative adaptive evolution

We also doubt that replicators are necessary for the cumulative evolution of complex features. Here is an example of a transmission system which does just that. When you speak, the kind of sounds that come out of your mouth depends on geometry of your vocal tract. For example, the consonant *p* in *spit* is created by momentarily bringing your lips together with the glottis open. Narrowing the glottis converts this consonant to *b* as in *bib*. Leaving the glottis open and slightly opening the lips produces *pf* as in the German word *apfel* (apple). Linguists have shown that even within a single speech community individuals vary in the exact geometry of the vocal tract used to produce any given word. Thus, it seems plausible that individuals vary in the culturally acquired rule about how to arrange the inside of the mouth when they are speaking any particular word. Languages vary in the sounds used and this variation can be very long lived. For example, in dialects spoken in the northwest of Germany, *p* is substituted for *pf* in *apfel* and many similar words. This difference arose about AD 500 and has persisted ever since (Bynon 1977).

So how are different rules governing speech production transmitted from generation to generation? Consider two models:

First, suppose that each child learning language is exposed to the speech of a number of adults. These adults vary in the way that they produce the *pf* sound in *apfel*. Each child figures out how she would need to position her tongue to produce the same *pf* sound as each adult model, and then she adopts *one* of these as her own rule. Here, a mental rule that governs speech production is transmitted from one individual to another. The mental rule is a replicator; it clearly has fidelity. It has longevity because it potentially persists for generations, and it would have fecundity if the rule was more attractive that competing rules. And because it is a replicator, it can evolve.

Now consider a second model. As before, children are exposed to the speech of a number of adults who vary in the way that they pronounce *pf*. Each child unconsciously computes the average of all the pronunciations that he hears and adopts the tongue position that produces this average. Here, mental rules are not transferred from one brain to another. The child may adopt a rule that is unlike any of the rules in the brains of its models. The rules in particular brains do not replicate because no rule is copied faithfully. The phonological system can nonetheless evolve in a quite Darwinian way. More attractive forms of pronunciation can increase if they have a disproportionate effect on the average. Rules affecting different aspects of pronunciation can recombine and thus lead to the cumulative evolution of complex phonological rules. It is true that the act of averaging will tend to decrease the amount of variation in the population each generation. However, phenotypic performances will vary as a result of age, social context, vocal tract anatomy, and soon. Learners will often misperceive an utterance. These sorts of errors in transmission will keep pumping variation into a population as averaging bleeds it away. In fact, averaging might be necessary to prevent high noise levels from injecting too much variation into the population (see Cavalli-Sforza and Feldman 1981; Boyd and Richerson 1985).

There are still other possibilities that differ even more radically from the replicator model. For example, a propensity to imitate the common type in the population can be coupled with high rates of individual learning to create a model in which there is little heritable variation at the individual level, but substantial heritability of group differences (Henrich and Boyd 1998). In such a model the cumulative evolution of adaptive complexity can occur, and occur rapidly, through selective processes that act at the group level (Boyd and Richerson 1990, in press). Similarly, in recent models of the evolution of social institutions (Young 1998), there is no cultural transmission at the individual level. Although individuals simply acquire the best response to their social environment by trial and error learning, the structure of social interactions creates persistent, heritable variation at the group level.

We do not understand in detail how culture is stored and transmitted, so we do not know whether culturally transmitted ideas and beliefs are replicators or not. If the application of Darwinian thinking to understanding cultural change depended on the existence of replicators, we would be in trouble. Fortunately, culture need not be closely analogous to genes. Ideas must be gene-like to the extent that they are somehow capable of carrying the cultural information necessary to give rise to the cumulative evolution of complex cultural patterns that differentiate human groups. They exhibit the essential Darwinian properties of fidelity, fecundity, and longevity, but, as the example of phonemes shows, this can be accomplished by a most ungene like, replicatorless process of error-prone phenotypic imitation. All that is really required is that culture constitutes a system maintaining heritable variation.

Darwinian models are useful

Science on the frontier often has an anarchic, nervy flavor because it must deal with multiple uncertainties. Of course, we would be better off knowing exactly what memes are. Papering over the uncertainties of how culture is stored and transmitted no doubt leads to errors, and conceals areas of fruitful inquiry. But as the psychologists explore one part of the frontier, the evolutionists should probe others. Studying the population properties of cultural information has lots of implications for human cognitive psychology, and vice versa. For example, when a child has the chance to copy the behavior of several different people, does she choose a single model for a given, discrete cultural attribute? Or, does she average, or in some other way combine, the attributes of alternative models? The minute you try to build a population model of culture you see that this question is crucial. However, despite conducting thousands of experiments on social learning, psychologists apparently have never thought to answer this question. Just as at a four-way stop, it makes no sense for everyone to wait for everyone else. Watch what the other drivers are doing, certainly, but go whenever the road ahead is clear.

Many social scientists have reacted to the advent of Darwinian models of culture with palpable distaste (e.g. Hallpike 1986), while others have embraced these ideas with enthusiasm (e.g. Runciman 1998). Much of this variation can be explained by people's feelings about the current Balkanization of the social sciences. The world of social science is divided into self-sufficient 'ethnies' like anthropology and economics that are content to follow the questions and presuppositions that govern their discipline. The inhabitants of this world regard other disciplines with a mixture of fear and contempt, and take little interest in what they have to say about questions of mutual interest. Clearly, this is not a satisfactory state of affairs.

We believe that Darwinian models can help rectify this problem. Disciplines, such as economics, psychology, and evolutionary biology, take the individual as the fundamental unit of analysis. These disciplines differ about how to model the individual and her psychology, but because they have the same fundamental structure there has been much substantive interaction between them. Nowadays, many economists and psychologists work closely together, and a rich new body of work, often called 'behavioral economics', has rapidly become mature enough to be applied to important practical problems such as the effect of retirement accounts on national savings rates. In the same way, economists and evolutionary biologists have found it relatively easy to work together on evolutionary models of social behavior, a rapidly growing field in both disciplines. Other disciplines like cultural anthropology and sociology emphasize the role of culture and social institutions in shaping behavior, and researchers in sociology, anthropology, and history find interaction with each other relatively comfortable. Bridging the gap between the individual and cultural disciplines has proved much more difficult. Darwinian models are useful precisely because they incorporate both points of view within a single theoretical framework in which individuals and culture are articulated in a way that captures some, if not all, of the properties that their respective specialists claim for them. In population-based models, culture and social institutions arise from the interaction of individuals whose psychology has been shaped by their social milieu. As a bonus, Darwinian models come with tools to investigate the population-wide, long-term consequences of the interactions between individuals and their culture and social institutions.

To see how useful population-based models can be, consider the problem of human cooperation. There is no coherent explanation for the vast scale of cooperation in contemporary human societies, or why the scale of cooperation has increased many 1000-fold over the last 10 000 years. Models in economics and evolutionary biology predict that cooperation should be limited to small groups of relatives and reciprocators. Many theories in anthropology simply assume (often implicitly) that cooperative societies are possible, and that culturally transmitted beliefs and social institutions serve the interest of social groups, but no attempt is made to reconcile this assumption with the fact that people are at least partly self-interested. Darwinian models provide one cogent mechanism to explain human cooperation by identifying the conditions under which groups will come to vary culturally, and predicting when such variation will lead to the spread of culturally transmitted beliefs that support large scale cooperation (Soltis *et al.* 1995). In such models, the effect of different culturally transmitted beliefs on group prestige and group survival shapes the kinds of beliefs that survive and spread. These group-level effects in turn influence what people want and what they believe, and therefore their behavior. Other recent work on the evolution of institutions (Young 1998; Richerson and Boyd in press) makes us optimistic that Darwinian models may have widespread utility.

Population thinking is also useful because it offers away to build mathematical theory of human behavior that captures the important role of culture in human affairs. Mathematical theory has the great advantage of allowing conclusions to be reliably deduced from assumptions. Experience in economics and evolutionary biology also suggests that it leads to a kind of clear understanding that is difficult to achieve with verbal reasoning alone. Of course there is also a cost—mathematical theory is necessarily based on simplified models. However, the combination of mathematical and verbal reasoning is superior to either alone.

Memes are not a universal acid, but population thinking is a better mousetrap. Population modeling of culture offers social science useful conceptual tools, and handy mathematical machinery that will help solve important, long-standing problems. It is not a substitute for rational actor models, or careful historical analysis. But it is an invaluable complement to these forms of analysis that will enrich the social sciences.

References

Bandura, A. (1986). *Social foundations of thought and action: A social cognitive theory*. Englewood Cliffs, NJ: Prentice-Hall.

Blurton, Jones, N. and Konner, M. J. (1976). !Kung knowledge of animal behavior. In *Kalahari hunter-gatherers: Studies of the !Kung San and their neighbors* (ed. R. Lee and I. DeVore), Cambridge, MA: Harvard University Press.

Boyd, R. and Richerson, P. J. (1985). *Culture and the evolutionary process*. Chicago IL: University of Chicago Press.

Boyd, R. and Richerson, P. J. (1990). Group selection among alternative evolutionarily stable strategies. *Journal of Theoretical Biology*, **145**: 331–42.

Boyd, R. and Richerson, P. J. (in press). Norms and bounded rationality. In *The adaptive tool box* (ed. G. Gigerenzer and R. Selten), Cambridge, MA: MIT Press. [Preprint available on the Web at: http: //www.sscnet.ucla.edu/anthro/faculty/boyd/]

Boyer, P. (1994). *The naturalness of religious ideas: A cognitive theory of religion*. Berkeley: University of California Press.

Bynon, T. (1977). *Historical linguistics*. Cambridge: Cambridge University Press.

Cavalli-Sforza, L. L. and Feldman, M. (1981). *Cultural transmission and evolution*. Princeton: University Press, Princeton.

Dawkins, R. (1976). *The selfish gene*. Oxford: Oxford University Press.

Dawkins, R. (1982). *The extended phenotype*. Oxford: Oxford University Press.

Dennett, D. (1995). *Darwin's dangerous idea*. London: Penguin.

Galef, B. G. (1988). Imitation in animals: History, definitions, and interpretation of data from the psychological laboratory. In *Social learning, psychological and biological perspectives* (ed. T. Zentall and B. G.Galef, Jr.), pp. 3–29. Hillsdale, NJ: Erlbaum.

Henrich, J. and R. Boyd. (1998). The evolution of conformist transmission and the emergence of between-group differences. *Evolution and Human Behavior*, 19: 215–42.

Hallpike, C. R. (1986). *The principles of social evolution*. Oxford: Clarendon Press.

Levebre, L. and Palameta, B. (1988). Mechanisms, ecology, and population diffusion of socially-learned, food-finding behavior in feral pigeons. In *Social learning, psychological and biological perspectives* (ed. T. Zetall and B. G.Galef, Jr.), pp. 141–65. Hillsdale, NJ: Erlbaum.

Mayr, E. (1982). *The growth of biological thought*. Cambridge, MA: Harvard University Press.

Needham, J. (1978). *The shorter science and civilisation in China* (Vol. 1). Cambridge: University Press, Cambridge.

Richerson, P. J. and R. Boyd. (1999). Complex societies: The evolutionary origins of a crude superorganism. *Human Nature*. 10: 253–89.

Runciman, W. G. (1998). Greek hoplites, warrior culture, and indirect bias. *The Journal of the Royal Anthropological Institute*, 4: 731–51.

Salomon (1992). *Prairie patrimony: Family, farming, and community in the midwest*. Chapel Hill: University of North Carolina Press.

Soltis, J., Boyd, R. and Richerson, P. J. (1995). Can group functional behaviors evolve by cultural group selection? An empirical test. *Current Anthropology*. 36: 473–94.

Tomasello, M., Kruger, A. C. and Ratner, H. H. (1993). Cultural learning. *Behavioral and Brain Sciences*, 16: 495–552.

Visalberghi, E. and Fragazy, D. M. (1990). Do monkeys ape? In *Language and intelligence in monkeys and apes* (ed. S. Parker and K. Gibson), pp. 247–73. Cambridge: University Press, Cambridge.

Whiten, A. and Ham, R. (1992). On the nature and evolution of imitation in the animal kingdom: A reappraisal of a century of research. In *Advances in the study of behavior*, Vol. 21 (ed. P. J. B. Slater, J. S. Rosenblatt, C. Beer, and M. Milkinski), pp. 239–83. Academic Press, New York.

Young, H. P. (1998). *Individual strategy and social structure: An evolutionary theory of institutions*, Princeton: Princeton University Press.

CHAPTER 8

An objection to the memetic approach to culture

Dan Sperber

Memetics is one possible evolutionary approach to the study of culture. Boyd and Richerson's models (1985, Boyd this volume), or my epidemiology of representations (1985, 1996), are among other possible evolutionary approaches inspired in various ways by Darwin. Memetics however, is, by its very simplicity, particularly attractive.

The memetic approach is based on the claim that culture is made of memes. If one takes the notion of a meme in the strong sense intended by Richard Dawkins (1976, 1982), this is indeed an interesting and challenging claim. On the other hand, if one were to define 'meme', as does the Oxford English Dictionary, as 'an element of culture that may be considered to be passed on by non-genetic means', then the claim that culture is made of memes would be a mere rewording of a most common idea: anthropologists have always considered culture as that which is transmitted in a human group by non-genetic means.

Richard Dawkins defines 'memes' as cultural replicators propagated through imitation, undergoing a process of selection, and standing to be selected not because they benefit their human carriers, but because they benefit themselves. Are non-biological replicators such as memes theoretically possible? Yes, surely. The very idea of non-biological replicators, and the argument that the Darwinian model of selection is not limited to the strictly biological are already, by themselves, of theoretical interest. This would be so even if, actually, there were no memes. Anyhow, there are clear cases of actual memes, though much fewer than is often thought. Chain-letters, for instance, fit the definition. The very content of these letters, with threats to those who ignore them and promises to those who copy and send them, contributes to their being copied and sent again and again. Chain-letters do not benefit the people who copy them, they benefit their own propagation. Moreover, some chain-letters are doing better than others because of the greater effectiveness of their content in causing replication.

Once the general idea of a meme is understood—and especially if it understood fairly loosely—it is all too easy to see human social life as teeming with memes. Aren't, for instance, religious ideas, with their threats of hell for unbelievers and promises of paradise for the proselytes, comparable to chain-letters, and in fact much more effective in benefiting their own propagation, come what may to their human carriers? More generally, aren't words, songs, fashions, political ideals, cooking recipes, ethnic prejudices, folktales, and just about everything cultural, items that get copied again and again, with the more successful items managing to invade more minds over longer periods of historical time, and to recruit those minds to further their own propagation? If this were so, if culture were made of memes in Dawkins's strong sense, then the study of culture could—and arguably should—be recast as a science of memes or 'memetics'. The Darwinian model of selection could be used, with proper adjustments, to explain the properties, the variety and the evolution of culture, just as it explains the properties, the variety, and the evolution of life.

The question is whether the claim that culture is made of memes is a true one. Several objections have been made to this claim. In his 'Foreword' to Susan Blackmore's *The Meme Machine* (1999), Richard Dawkins responds to the simplest and most serious objection, 'that memes, if they exist at all, are transmitted with too low fidelity to perform a gene-like role in any realistically Darwinian selection process' (Dawkins 1999: x).[1] I want here to discuss Dawkins's responses, and, in so doing, develop a different fundamental objection to the meme model. This new objection is that most cultural items are 're-produced' in the sense that they are produced again and again—with, of course, a causal link between all these productions—but are not reproduced in the sense of being copied from one another (see also Origgi and Sperber,

---

[1] Dawkins adds: 'The difference between high fidelity genes and low fidelity memes is assumed to follow from the fact that genes, but not memes, are digital'. The objection that memes are transmitted with too low fidelity can be made without this further claim, which I find vague and uncompelling.

forthcoming). Hence they are not memes, even when they are close 'copies' of one another (in a loose sense of 'copy', of course).

The objection of low fidelity had been envisaged and taken seriously by Dawkins himself. In *The Extended Phenotype* (Dawkins 1982: 112) he wrote:

The copying process is probably much less precise than in the case of genes: there may be a certain 'mutational' element in every copying event [...]. Memes may partially blend with each other in a way that genes do not. New 'mutations' may be 'directed' rather than random with respect to evolutionary trends. [...] there may be 'Lamarckian' causal arrows leading from phenotype to replicator, as well as the other way around. These differences may prove sufficient to render the analogy with genetic natural selection worthless or even positively misleading. My own feeling is that its main value may lie not so much in helping us to understand human culture as in sharpening our perception of genetic natural selection.

Of course, what counts as 'too low fidelity' for a given item is relative to the selection bias for that item (see Williams 1966). A greater selection bias allows for a higher mutation rate. On the other hand if, as Dawkins says, there is 'a certain "mutational" element in every copying event' (loc. cit.), then it is not easy to see how selection could work at all. It is to this problem that Dawkins (1999) now offers an ingenious solution. He uses for this a thought experiment of which I present a simpler but equally effective version (before discussing his version later). Consider Figure 8.1. A first individual is shown this figure for ten seconds and is asked, ten minutes later to reproduce it as exactly as possible. Then a second individual is shown for ten seconds the figure drawn by the first individual and presented with the same task. This is iterated with, say,

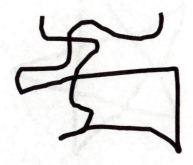

Figure 8.1

nine participants. It is most likely that each drawing will differ from its model and that the more distant two drawings are in the chain, the more they are likely to differ. A judge given the ten drawings in a random order and asked to put them back in the order in which they were produced should perform, if not perfectly, at least much better than random. The 'mutational elements' in every copying event are such that a drift is manifest, and no stable pattern is maintained.

Now imagine a similar experiment being performed, but this time with Figure 8.2 as initial input. Again, each drawing produced by the successive participants is certain to differ from its model, since each participant will fail to reproduce the model in all its particulars. This time, however, the distance in the chain of two drawings on the one hand, and their degree of difference on the other hand should be two variables independent of one another (or nearly so). A judge asked to put the ten drawings in the order in which they were produced should be unable to do better than random. Despite low fidelity of copying, a stable pattern is most likely to endure across versions, and individual variations are very unlikely to compromise this pattern.

What explains the difference between the two experiments? In the case of Figure 8.1, people try and form a mental image of a drawing which they do not recognize in any way, and then try and reproduce this mental image on paper. In storing the information, in recalling it, and in reproducing it, they are likely to introduce unintended variations that are either in random directions, or are in the direction of entropy, that is, plain loss of information. In the case of Figure 8.2,

Figure 8.2

people recognize the figure as a five-branched star drawn without lifting the pencil. They may well forget most of the other particulars of the drawing under their eyes, such as length of relatively straight segments, or angles. Still, they will produce another star of the same type.

Dawkins might describe the difference between the two types of tasks as follows. In tasks of the first type, what gets copied is the product, the drawing. There is no difference therefore, between the 'phenotype' and the 'genotype', and phenotypic variations are also genotypic variations. In cases of the second type, what gets copied is the implicit instruction ('draw a five-branched star without lifting the pen'). These instructions are the true genotype, while the drawings are only phenotypes. Each participant in the experiment assumes that the preceding participant merely intended to follow the implicit instruction, and that imperfections or idiosyncrasies were unintended and should be ignored. Individual variations in the productions of the phenotype do not matter. They are not genuine mutations. 'The instructions,' writes Dawkins 'are self-normalising. The code is error-correcting' (1999: xii).

Dawkins concludes the argument by stating: 'I believe that these considerations greatly reduce, and probably remove altogether, the objection that memes are copied with insufficient high fidelity to be compared with genes. For me the quasi-genetic inheritance of language, and of religious and traditional customs, teaches the same lesson' (p. xii). In other words, the stability of cultural patterns is proof that fidelity in copying is high despite individual variations. These variations are phenotypic, not genotypic, and Darwinian selection can take place without being jeopardized by too high a rate of mutation.

I, on the other hand, believe that what is here offered as an explanation is precisely what needs to be explained; what is offered as a solution is in fact the very problem to be solved. Saying that the instructions are 'self-normalising' amounts to resolving a problem by invoking a mystery. The type of thought experiment proposed by Dawkins is well worth analysing so as to solve the mystery. The conclusions I draw from this thought experiment are, however, very different from that of Dawkins. They point to yet another difficulty with the meme model.

Let me grant forthwith two points to Dawkins:

1. Of course, one item A can be a replica (in the relevant sense) of another item B without being identical to B in every respect. From a memetic point of view, it is enough that A and B should share the properties the recurrence of which one is trying to explain.

2. Of course, cultural items exhibit, over periods of time of various length (longer for folktales, shorter for modern dress fashions, for instance), the kind of stability found, on a much smaller scale, in Dawkins thought experiment. That is, although there is much individual variation, items of the same type all remain in the vicinity of one another and instantiate a common pattern.

The issue is whether the relative stability found in cultural transmission is proof of replication. Dawkins seems to think it is. In substance, he proposes a test to decide whether a causal chain that links the production of a series of items is a chain of replications. The test is as follows. Present (or suppose you present) to an intelligent observer the items in the chain in a random order. If the observer finds it impossible to put back, at least approximately, the items in the order in which they were produced, then these items are replications in the relevant sense. Individual variations among these items are phenotypic and do not compromise the stability of the underlying genotype. Much of culture passes this test and is seen, then, as made of replicators.

To show that Dawkins test is not as reliable as it may seem, let me first give an example of a causal chain that would meet the criterion, but could not be properly described as a case of memetic transmission. Consider the case of laughter. Laughter is a social behaviour that is typically triggered, in individual development, by the laughter of others, and that remains a highly contagious form of behaviour. Laughter is influenced in its intensity, style, and circumstances of arousal by cultural factors. Moreover, even within a cultural group, there are important individual variations. Now, imagine a series of registerings of causally linked individual laughters (linked either in the stabilization of laughing behaviour across generations, or in a much shorter causal chain of contagious laughter). If these registerings were presented in a random order, they could not, I take it, be rearranged in their causal order. Laughter passes Dawkins test. Yet, surely, it is not a meme.

Why is laughter not a meme? Because it is not copied. A young child who starts laughing does not replicate the laughters she observes. Rather, there is a biological disposition to laughter that gets activated and fine-tuned through encounters with the laughter of others. Similarly, an individual pushed into convulsive laughter by the laughter of others is not imitating them. The motor program for laughing was already fully present in him, and what the laughter of others does is just activate it.

Let me generalize and define three minimal conditions for true replication. For B to be a replication of A,

(1) B must be caused by A (together with background conditions),

(2) B must be similar in relevant respects to A, and

(3) The process that generates B must obtain the information that makes B similar to A from A.

Another way to express this third condition is to say that B must inherit from A the properties that make it relevantly similar to A. Discussions of memes take implicitly for granted that the co-occurrence of causation and of similarity between cause and effect is sufficient evidence of inheritance. But this is not so. The cause may merely trigger the production of a similar effect, as we saw with the case of laughter. Even if conditions (1) and (2) are satisfied, condition (3) may not be.

Consider a theoretical example, with two cases to be compared. In both cases conditions (1) and (2) are satisfied, but condition (3) is satisfied only in the second case. First case: ten sound-recorders with the same repertoire of melodies in each have been fixed so that they are activated by the sound of the last five bars of any melody in their repertoire, and then play this very melody. They are placed in such a manner and at such a distance of one another that the first one activates the second, the second the third, etc. The first recorder plays melodies in random order at appropriate time intervals. Second case: ten sound-recorders have been fixed and placed so that the second-recorder records sound from the first, and then replays it, the third recorder records sound from the second and them replays it, and so on. Only the first recorder has a ready repertoire of melodies, and it plays them in random order at appropriate time intervals. In both cases, an observer listening to these devices playing, each in turn, one melody after another, and unable to inspect them otherwise, would have some reasons to think she was witnessing a series of replications. In fact, this would be true in the second case, but not in the first, where only triggering takes place and no copying at all. This illustrates the point that, in the case of a causal chain that satisfies conditions (1) and (2), further evidence about the causal processes involved must be available before one is in a position to argue that condition (3) is also satisfied, and that one is dealing, therefore, with a true chain of replications.

Let us go back, now, to our thought experiment. In the first task (memorizing and reproducing Figure 8.1), participants rely on general perceptual, memory and motor abilities. In other words, they rely on the general human ability to imitate, an ability which is taken by memeticists to be extremely powerful. In this case, however, it fails. In the second task (memorizing and reproducing Figure 8.2), the stimulus is recognized. That is, it triggers the activation of pre-existing knowledge. The stimulus is categorized as a token of a general type: a five-branched star drawn without lifting the pencil. Properties of the actual stimulus that are irrelevant to this categorization are just ignored. When asked, after ten minutes, to reproduce the stimulus, participants just produce another token of a five-branched star without, in most cases, even trying to remember what the original figure exactly looked like. Their ability to perform well in this second task is not an ability to perceive and copy. It is an ability to recognize and re-produce, using, for this, knowledge of the five-branched star type that they already possessed before encountering the token. It is not, then, that people are better at imitating Figure 8.1 than at imitating Figure 8.2. They are indeed bad at imitating Figure 8.1, and they are not imitating Figure 8.2 but merely producing a new token of the same recognizable type.

Dawkins's original thought experiment involved a comparison of two tasks: reproducing a drawing of a Chinese junk, or making an origami Chinese junk after having been taught, by demonstration, how to make one. Unlike my simpler version, the two final products—the drawing or the origami—are recognized by the participants. In the drawing version, however, participants are unable to recognize the series of strokes that would yield the full drawing, whereas in the origami version the successive foldings are individually demonstrated. Thus, the two task are different, not just in the type of item to be copied (a drawing vs. an origami) but also in the fact that participants observe only the product in the first task, and the process of production in the second task. If participants were just shown a finished origami junk, they would, presumably, do even worse in reproducing it than in reproducing a drawing of a junk.

The crucial difference between the two tasks is that the second involves demonstration, and the other not. From the demonstration, or so Dawkins assumes, participants can and do infer implicit instructions (e.g. 'take a square sheet of paper and fold all four corners exactly into the middle'). These instructions are not a description of what the person making the origami is actually doing (the four corners are never folded exactly into the middle, for instance) but a description of what the person is aiming at, is intending to do. Inferring instruction involves much more than the ability to perceive and describe actual movements; it involves the ability to attribute goals and intentions.

Contrary to what Dawkins writes, the instructions are not 'self-normalizing'. It is the process of attribution of intentions that normalizes the implicit instructions that participants infer from what they observe. When you see the person folding the four corners of a square sheet of paper into four different points in the vicinity of the middle, you assume that she was aiming at the middle rather than at these four odd points. Such intentions to realize regular geometrical pattern are familiar—in particular, in the context of origami—and readily attributed. You recognize, in other terms, the behaviour as an imperfect realization of an intention of a familiar and regular type rather than as the perfect realization of an intention of an unfamiliar and irregular type. The instructions that you infer are, then, informed in part by what you actually observe, and in part by what you already know of human intentions, and of the type of instructions typically used in origami.

The instructions are not being 'copied' in any useful sense of the term from one participant to the next. Certainly, instructions cannot be imitated, since only what can be perceived can be imitated. When they are given implicitly, instructions must be inferred. When they are given verbally, instructions must be comprehended, a process that involves a mix of decoding and inference (Sperber and Wilson 1995). The inference involved in either case draws on domain-specific competencies having to do with the attribution of intentions and with knowledge of the role of regular geometric forms in the formation of human intentions generally, and in paper-folding in particular. Thus, the normalization of the instructions results precisely from the fact that something other than copying is taking place. It results from the fact that the information provided by the stimulus is complemented with information already available in the system.

In the real world, and in particular in the cultural world, triggering and copying can and do combine in various degrees. What gets triggered by cultural stimuli are acquisition mechanisms and competencies that are more or less domain-specific. These mechanisms are themselves in part genetically, in part culturally, inherited.

Let us briefly consider the example of the acquisition of language. In acquiring a language, a child internalizes a grammar and a lexicon on the basis of linguistic interactions. Nowhere in these interactions—nowhere in the linguistic data the child is presented with—is the grammar present to be copied. Rather, the grammar must be inferred from these data. As Noam Chomsky has long argued and as has become, if not universally, at least generally accepted today, this requires a genetically determined preparedness to interpret the data in a domain-specific way and to generalize from it to the grammar of the language, going well beyond the information given. Imitation in some sense may well play a role—though not a sufficient one—in the acquisition of the phonology of words, but not in the acquisition of their meaning. Meaning is not something that can be observed and copied. It can only be inferred. Language learners converge on similar meanings on the basis of weak evidence provided by words used in an endless diversity of contexts and with various degree of literalness or figurativeness. Acquisition of meaning in such conditions is a feat that would be wholly mysterious if it were not highly constrained by domain-specific competencies having to do with conceptual domains on the one hand, and with the attribution of communicative intentions to speakers on the other. Thus, the similarities between the grammar and lexicons internalized by different members of the same linguistic community owe little to copying and a lot to pre-existing linguistic, communicative, and conceptual evolved dispositions.

The respective role of copying and that of pre-existing dispositions to construe evidence in domain-specific structured ways may vary with different cultural competencies. Learning to tap-dance involves more copying than learning to walk. Learning poetry involves more copying than learning philosophy. For memetics to be a reasonable research programme, it should be the case that copying, and differential success in causing the multiplication of copies, overwhelmingly plays the major role in shaping all or at least most of the contents of culture. Evolved domain-specific psychological dispositions, if there are any, should be at most a relatively minor factor that could be considered part of background conditions. There is nothing obvious about such a view. While the view may have some popularity among unconcerned lay people, no psychologist believes that cultural learning is essentially a matter of imitation (this is true even of psychologists who attribute an important role to imitation, e.g., Meltzoff and Gopnik 1993; Tomasello et al. 1993). In fact, such an idea goes against all major recent developments in developmental psychology and in evolutionary psychology (see Hirschfeld and Gelman 1994). This, together with the problem raised in this article, puts a special burden on memeticists.

Memeticists have to give empirical evidence to support the claim that, in the micro-processes of cultural transmission, elements of culture inherit all or nearly all their relevant properties from other elements of culture that they replicate (i.e. satisfy condition 3 above). If they succeeded in doing so they would have shown that developmental psychologists, evolutionary psychologists, and cognitive anthropologists who argue that acquisition of cultural knowledge and know-how is made possible and partly shaped by evolved domain-specific competencies are missing a much simpler explanation of cultural learning: imitation does it all (or nearly so)! If, as I believe, this is not even remotely the case, what remains of the memetic programme? The idea of a meme is a theoretically interesting one. It may still have, or suggest, some empirical applications. The Darwinian model of selection is illuminating, and in several ways, for thinking about culture. Imitation, even if not ubiquitous, is of course well worth investigating. The grand project of memetics, on the other hand, is misguided.

References

Boyd, R. and Richerson, P. J. (1985). *Culture and the evolutionary process.* Chicago: The University of Chicago Press.

Dawkins, R. (1976). *The selfish gene.* Oxford: Oxford University Press.

Dawkins, R. (1982). *The extended phenotype.* Oxford: Oxford University Press.

Dawkins, R. (1999). Foreword *The meme machine* by Susan Blackmore. Oxford: Oxford University Press.

Hirschfeld, L. and Gelman, S. eds. (1994). *Mapping the mind: Domain specificity in cognition and culture.* New York: Cambridge University Press.

Origgi, G. and Sperber, D. (forthcoming). Evolution, communication, and the proper function of language. In *Evolution and the human mind: Language, modularity, and social cognition* (ed. P. Carruthers and A. Chamberlain). Cambridge: Cambridge University Press.

Meltzoff, A. and Gopnik, A. (1993). The role of imitation in understanding persons and developing a theory of mind (ed. S. Baron-Cohen *et al.*), *Understanding other minds.* Oxford: Oxford University Press.

Sperber, D. (1985). Anthropology and psychology: towards an epidemiology of representations. *Man* (N.S.), **20**: 73–89.

Sperber, D. (1996). *Explaining culture: A naturalistic approach.* Oxford: Blackwell.

Sperber, D. and Wilson, D. (1995). *Relevance: Communication and cognition* (2nd edn). Oxford: Blackwell.

Tomasello, M., Kruger, A. and Ratner. H. (1993). Cultural learning. *Behavioral and Brain Sciences,* **16**: 495–552.

Williams, G. C. (1966). *Adaptation and natural selection.* Princeton: Princeton University Press.

CHAPTER 9

If memes are the answer, what is the question?

Adam Kuper

If memes are the answer, what is the question? The question that memes are designed to address evidently concerns culture, but culture is itself a notoriously question-begging notion. And culture is supposed to provide the answers to another very big question, which is in what way human beings may be unique.

'Most of what is unusual about man can be summed up in one word: "culture"', Dawkins wrote, continuing, with a perhaps disingenuous insouciance, 'I use the word not in its snobbish sense, but as a scientist uses it' (Dawkins 1989: 189). Unfortunately, he does not specify how a scientist uses the word, and little wonder. In truth, there is no single, unsnobbish, scientific conception of culture.[1]

What Dawkins refers to as the snobbish idea of culture was most famously summed up in an aphorism of Matthew Arnold: culture is the best that has been thought and said. It is the sum of the greatest spiritual and artistic accomplishments of humanity (which meant the finest flower of the high art of Europe). Culture marked the elect off from the masses, the civilized from the unlettered barbarians.

In 1871, two years after the appearance of Matthew Arnold's *Culture and Anarchy*, Darwin published *The Descent of Man*, which raised the question of what distinguished humans from other primates. In the same year, in a book provocatively entitled *Primitive Culture*, the pioneer anthropologist E. B. Tylor answered that it was culture or civilization

---

[1] See Adam Kuper (1999).

that guaranteed the uniqueness of human beings. But it was not Matthew Arnold's culture that Tylor had in mind. For Matthew Arnold, culture distinguished the elite from the hoi-poloi. For Tylor, culture marked off humans from other primates. Tylor's culture was therefore not restricted to the elect, and it was not only a matter of high art. It was shared by all people, and it included every custom and skill that was transmitted by society rather than by biology, by nurture rather than nature. Every people, and every individual in every society, had culture. This common culture was, moreover, supposed to be constantly advancing, onwards and upwards. Just as (Tylor thought) human beings are clearly an advance on other primates, so human civilization gradually became better and better. In short, human history was the story of the progressive development of human culture.

This conception of culture or civilization was not altogether new. It was a modernized version of the established Enlightenment—or French—conception of the course of human history. In the French tradition, civilization was represented as a progressive, cumulative human achievement. The progress of civilization could be measured by the advance of reason in its cosmic battle against raw nature, instinct, and unthinking tradition. This advance was most obvious in science and technology, and in the growing rationality of government. Civilization had progressed furthest, of course, in France, but it was also enjoyed, if in different degrees, by savages, barbarians and other Europeans.

This Enlightenment conception of a common, progressive human civilization had been challenged almost from the first by what is sometimes referred to as a Counter-Enlightenment movement, which became established especially in intellectual circles in Germany. Following Herder, it insisted on the differences between populations, and argued that these differences were essentially cultural. *Kultur*, moreover, was associated with spiritual rather than material values. Its affinities were with religion, and its most characteristic achievements were to be found in the arts rather than the sciences. Each *Volk* had its own *Geist*, and its particular spiritual values were expressed above all in its language and arts.

To sum up, in the French tradition, and in Tylor's anthropological formulation, culture or civilization was universal and progressive, and its central elements were science and technology. In the German tradition, a culture was the heritage of a particular community, and its culture distinguished each community from its neighbours. At its core was religion, language, and the arts.

Modern anthropology inherited both these conceptions of culture. For much of this century, American anthropology has been divided into two rival camps, one which continues the tradition of French positivism, the other that of German idealism. The one camp presents itself as 'evolutionist' and scientific. It treats a culture as essentially a machine for living, a set of tools for the exploitation of nature. The other camp is relativist, and defines culture as a system of ideas and values, expressed in symbols, characteristic of a particular population. For the first group, culture is what distinguishes us from the animals, and it is progressive. (Despite their claims to be the heirs of Darwin, these 'evolutionists' were generally firm believers in unilineal progress.) For the second group, culture is the specific world view that distinguishes one human population from another. There is no objective measure of cultural superiority. (On the other hand, every group believes itself to be uniquely excellent.) For those who described themselves as evolutionists, culture must satisfy natural needs. For the relativists, needs are culturally constructed and themselves therefore culturally variable.

Culture and progress

Dawkins never seems to cite contemporary anthropological writers on culture, or classical writers for that matter. Nevertheless, I suspect that his ideas about culture are a throwback to an earlier and happier time. His closest affinity is perhaps with a particular faction of English Victorian 'evolutionists' which was led by E. B. Tylor. In this tradition, human culture is constituted largely by knowledge of nature, by the (consequent) ability to control nature, and by the progressive implementation of moral rules that suppress our own animal nature. This common culture is in the process of development. It is possible to be more or less civilized. Some nations or peoples are in the forefront of progress. Others lag far behind. (It seemed apparent to the Victorian anthropologists that tropical savages lived in much the same way as the earliest Europeans. 'The astonishment which I felt on first seeing a party of Fuegians on a wild and broken shore will never be forgotten by me', Darwin recalled, 'for the reflection at once rushed into my mind—such were our ancestors', (Darwin 1871: 919–20). The measure of progress was self-evident to Tylor and to Frazer and, it turns out, to Dawkins. The most primitive peoples still believe in religion, and try to manipulate nature by magical techniques. The most civilized put their faith in science and technology.

By what means did a people climb up the ladder of progress? For Darwin, the answer seemed evident. Just as human beings had larger brains than the apes, so more advanced humans had larger brains than primitive humans. And as their brains grew, so people advanced from a belief in magic to a faith in religion to a knowledge of science, from hunting and gathering like animals to a mastery of nature, from promiscuous coupling to faithful monogamous marriage.

Various arguments were advanced against this model within anthropology, but I shall simply pick out two of these. First, a school of anthropologists who came to be called diffusionists pointed out that people in a geographical region often shared many ideas and customs even if they apparently stood at different levels of evolution. In the Cape, for example, so-called Bushman hunters and gatherers had the same languages and religious ideas and marriage rules as the Hottentots, who were pastoralists. Second, the diffusionists argued that improved, more efficient techniques and practices were generally introduced by borrowing rather than by independent development. Indeed, people were often forced to change as a result of conquest, or they might react against the model that was imposed on them by more powerful, foreign invaders.

Later anthropologists also began to question the simple faith in progress that characterized this form of 'evolutionism'. Technology certainly becomes more and more efficient. Science advances on all fronts. However, while machines, communications, medicine, agriculture, etc. may be based on the most up-to-date scientific ideas, this does not mean that a modern society is in some general sense rational or scientific. It is probably the case that the great majority of people in the most sophisticated circles in Europe and North America could not give an accurate account of the most basic ideas in modern physics or biology. Moreover, advances in science and technology do not necessarily undermine religious ideas within a society. The United States is at once the society with the most advanced technology and science in the world, and the most religious population in the West. In any case, although it may be easy enough to establish certain measures of progress in the fields of science and technology it is very difficult, if not impossible, to agree on equivalent measures of progress in other fields of human activity: morality, for example, or the arts, or kinship organization. (To be sure, all this is relevant to a critique of soft Lamarckian ideas rather than Darwinism. Darwinism is, of course, utterly opposed in principle to any teleological way of thinking. Nevertheless, faith in progress is probably one of the subliminal attractions of any 'evolutionary' theory of culture.)

The ecology of ideas

If Dawkins is somewhat cavalier in his treatment of cultural theories, he also takes it for granted that the facts about human culture are pretty straightforward, and do not require much in the way of investigation. For example, he invites us to consider the idea that there may be a meme for suicide. This is an interesting suggestion, since it seems unlikely that there can be a gene for suicide (though there may well be one for depression). Dawkins even suggests that memes drive suicide epidemics, arguing that 'a suicidal meme can spread, as when a dramatic and well-publicised martydom inspires others to die for a deeply loved cause' (Dawkins 1982: 111). In support, he cites Gore Vidal, 1955. This turns out to be an early novel by the American writer, about a messianic cult. Dawkins would surely be apoplectic if a social scientist were to cite Hitchcock's film *The Birds* to make a point about ornithology. There are, of course, well-documented and serious analytical accounts of group holocausts in various cult communities, many published since 1955, but Dawkins does not even gesture towards them. Had he consulted this literature he would have noted that it is by no means evident that all the victims of these so-called mass suicides killed themselves on their own volition, as willing sacrifices for a deeply loved cause. Were the children killed in Jonestown victims of their own memes? Any adequate account of these episodes has to take into account emotions as well as ideas, and power relations as well as individual replicators.

To take another example, in *The Selfish Gene* Dawkins asks why the meme for belief in god is so persistent (1989: 193), and offers the following analysis:

The survival value of the god meme in the meme pool results from its great psychological appeal. It provides a superficially plausible answer to deep and troubling questions about existence. It suggests that injustices in this world may be rectified in the next.

This is banal stuff, but while one might drowsily pass over it in a Sunday newspaper essay, presented in a scientific book it should be supported by some evidence. There is a large literature on religious belief, written by anthropologists, psychologists, historians, and other scholars. Dawkins cites none of it. Does the introduction of memes into the argument help us in any way to grasp why (some but not all) people believe in (very differently conceived) 'gods'? The conclusion that there is a meme for 'god' and that it survives because of its psychological appeal recalls the quack promoting a sleeping pill for its dormative qualities.

Even if memes are just ideas, and we specify the ideas rather more precisely than Dawkins has done in these instances, they should not then be treated as isolates. Unlike genes, cultural traits are not particulate. An idea about God cannot be separated from other ideas with which it is indissolubly linked in a particular religion. Judaeo-Christian monotheism is a very different system of ideas from Graeco-Roman polytheism, and both are very different in turn from Hindu polytheism. And ideas about gods are not free-floating, independent entities. The idea of a god is necessarily related to other cosmological ideas, such as creation, or incarnation, or reincarnation, or divine retribution. Some gods watch over us, others wash their hands of us. Further, a specific set of ideas about god usually operates within an institutional matrix, sustained by seminaries, prayer-books, holy days, rituals, churches, and so on. And these are in turn often bolstered by the power of the state. It is not easy, then, to separate out (for instance) a particular idea of god, and to assess its independent power.

Ethnographic studies of religious and ritual behaviour tend to show that the power of ideas and practices depends very often on their setting, in particular on the part they play in an intricate network of relationships between people, things, places, symbols, and other ideas. Consider an example from King's College, Cambridge. In the 1960s, an anthropologist, Edmund Leach, became provost of King's. He was a crusading atheist, and some optimists anticipated that he would abolish the religious side of college life. After all, chapel services were obviously incompatible with the scientific vocation. In the event, Leach did not turn the college chapel into a museum of dead religions. On the contrary, he came rather to enjoy the rituals in which he had to participate. But he remained an anthropologist, and he was struck by a curious fact about the central annual ritual of the college, Founder's Day.

Various ritual performances took place in different parts of the college in the course of the day, but Leach discovered that there was no master programme, and that nobody knew the whole Founder's Day ritual. He himself, as Provost, was instructed to do this and that as the day progressed by various temporary leaders of the ritual, but nobody could tell him what his duties were throughout the day. The college porters knew certain things that had to be done, some of the older fellows were expert in certain other parts of the performance, the chaplain could be relied upon to do his bit, as could the choirmaster, the chef had his notes on what had to be served at the Founder's feast, and so on.

If there was no single programme for the ritual, and no centre of ritual authority, it was unlikely that this complex performance expressed a single message, or embodied a straightforward idea. Leach decided to make an ethnographic study of the ritual. However, he was warned off by an influential group of college fellows. They wanted Founder's Day to remain a bit of a mystery. Perhaps that was, precisely, the idea of the ritual. It is also plausible that just because various constituencies within the college each controlled a part of the ritual, they were obliged to recognize their mutual dependence. For this reason it would be dangerous to publish the full script of the ritual, for all to see.

What are to be counted as memes in this package of rituals, relationships, communal meals, and choral performances? What (as one might say) is the big idea? Culture traits are not the equivalent of philosophical notions, and even when ideas are in question, their ecology is not made up only or even primarily by other ideas.

Culture versus genes

These are all queries about the utility of the idea of memes as a tool for cultural or social research. But this is perhaps to miss the real point of memes in the ecology of Dawkins' theory. Memes may have been specifically designed to torpedo the messianic human sociobiology of E. O. Wilson. Dawkins was writing a book on sociobiology, but he had no time for human sociobiology. He thinks that Wilson and his acolytes are quite wrong to regard human beings in the same way as other animals, or even birds or insects. 'Are there any good reasons for supposing our own species to be unique?' Dawkins asks, and answers that we are indeed unique, because we have culture (Dawkins 1989: 189). It is for this reason that we must exclude ourselves from the specific evolutionary arguments that apply to all other 'survival machines'. The Darwinians fell into error when they 'tried to look for "biological advantages" in various attributes of human civilization'. Dawkins concludes that 'for an understanding of the evolution of modern man, we must begin by throwing out the gene as the sole basis of our ideas on evolution' (Dawkins 1989: 191).

To be sure, the genes are not really thrown out. They remain, but in a new, ethereal role. 'The gene will enter my thesis as an analogy, nothing more' (Dawkins 1989: 191). The essential job that these gene analogies, the memes, perform is to push genes into the shadows, to reinforce the rather traditional belief that human beings are unique because they have ideas and ideals. 'My purpose was to cut the gene down to size,' Dawkins has explained, 'rather than to sculpt a grand theory of human culture' (Dawkins 1989: 323).

Dawkins sets up a rhyming opposition between memes and genes that recalls the old opposition between nature and nurture. In a familiar, indeed classical manner, Dawkins in effect splits human beings into two elements, higher and lower, spiritual and physical, mind and body. Our behaviour cannot be reduced to needs, instincts, or genes. Culture, nurture, consciousness, and, it now turns out, memes, allow us to tran-scend the animal state. Armed with memes, we can raise ourselves above our original condition. We may even learn to pick and choose between memes, using our reason to guide us. Like good scientists, we will consider the evidence and reject errant memes, especially those that carry religious ideas. (It is worth noting that a very high proportion of the examples of memes provided by Dawkins in *The Selfish Gene* are religious beliefs.) And we have free will. 'We are built as gene machines and cultured as meme machines, but we have the power to turn against our creators', Dawkins assures us. 'We, alone on earth, can rebel against the tyranny of the selfish replicators' (Dawkins 1989: 201).

But this rejection of human sociobiology leaves Dawkins with a large problem. If culture is not bound in the service of biology, does it follow that biologists have nothing to say about culture? If that is conceded, then biologists will have to accept that they cannot deliver a satisfac-tory theory of human behaviour. Perhaps they will have to go and learn some anthropology, or even, heaven help us, some sociology. Dawkins is not prepared to accept any such suggestion. Wilson may have erred in opting for genetic determism of culture, but Dawkins evidently agrees with him that a healthy dose of biology is needed to sort out sociology and psychology. However, Dawkins does not believe that human beings should be treated in the same way as ants or birds, as gene machines. What he prescribes is a dose of biological *theory*. The question then is, which biological theory will deliver a science of cultural evolution?

Other biologists have faced the same problem, repudiating human sociobiology but still convinced that somewhere in biology there must be a theory that will make sense of what happened after the human lineage split off from the other primates. Medawar and Gould, to take two distinguished examples, both rummaged through the musty attic of rejected biological ideas in order to find a second-hand notion that would fit the poor, deprived social sciences. They both came up with Lamarckism. 'Apart from being mediated through non-genetic channels', Medawar remarks, 'cultural inheritance is categorically distinguished from biological inheritance by being Lamarckian in character; that is to say, by the fact that what is learned in one generation may become part of the inheritance of the next' (Medawar 1982: 173). Gould came to the same conclusion: 'Human cultural evolution is Lamarckian—the useful discoveries of one generation are passed directly to offspring by writing, teaching, and so forth' (Gould, 1987: 70).

Like Medawar and Gould, Dawkins scorns human sociobiology. Like them again, he takes it for granted that the social sciences are sorely in need of a decent theory, and he assumes that a really good theory can only come from biology. He is also tempted by Lamarckism as a theory of culture (e.g., Dawkins 1982: 112), but in the end he opts for a neo-Darwinian trope. Hence, memes.

The genetic analogy

High concept sociobiology was a very hot topic some twenty years ago. Its leading entrepreneurs went around handing out promissory notes as though there was no tomorrow. It all seems very long ago now, and it is hard to find anyone who remembers ever arguing that genes cause cultural rules (such as the incest taboo) or practices (like male hunting or courtship dances). The ambition that drove sociobiology survives: to establish a Darwinian social science. However, as in advanced religious circles, literalism has given way to metaphorical readings. Genes do not literally programme cultural behaviour. Rather there is something about culture that is like something about genes. But what, precisely, is like what?

'Cultural transmission is analogous to genetic transmission', Richard Dawkins writes in *The Selfish Gene*, 'in that, although basically conservative, it can give rise to a form of evolution' (Dawkins revised edition 1989: 189). Cavalli-Sforza and Feldman suggest that the basic shared characteristic of learning and genetic transmission, from which all the rest flow, is that 'entities' can be passed from one person to another. Since 'copying' may give rise to errors, there is room for 'evolution'.[2] Boyd and Richerson (1985) agree on these very general propositions, but at the same time they insist on the *differences* between learning and the processes of genetic transmission. They draw on modern psychology to specify the very distinctive ways in which people learn (as Darwin had it, by way of 'habit, example, instruction, and reflection'). They then argue that learning combines with the distinct process of genetic transmission to constitute a 'dual inheritance' system that is uniquely human. In their view, the real analogy between cultural change and genetic evolution is to be found not in the process of replication but in the process of selection. Cavalli-Sforza and Feldman, however, explicitly contrast what they call 'cultural selection' with 'natural selection', and they insist that the two modes of 'selection' may be in tension with one another. And for Dawkins, natural selection seems to have little to do with the fate of memes. Their success rests simply on their ability to reproduce themselves. The mechanism of change is apparently memetic drift. (Can we now look forward to scientific memetic engineering?)

Metaphors may serve as a useful aid to clear thinking.[3] And it

[2] "Transmission may imply copying (or imitation); copying carries with it the chance of error. Thus we have in cultural transmission the analogs to reproduction and mutation in biological entities. Ideas, languages, values, behaviour, and technologies, when transmitted, undergo 'reproduction', and when there is a difference between the subsequently transmitted version of the original entity, and the original entity itself, 'mutation' has occurred. . . . Reproduction and mutation ensure that evolutionary change will take place." (Cavalli-Sforza and Feldman 1981: 10.) And having set up these analogies between cultural and genetic transmission, they proceed to apply mathematical models drawn from population genetics to instances of cultural change.

[3] There is, after all, the encouraging example of Darwin's famous bush. Darwin used the already established images of the bush or the tree to represent the genetic relationship between different populations, sketching a tree in the 1837 notebook, and elaborating the image in a rather more abstract fashion in his famous branching diagram, the sole visual aid to break the austere prose in the whole of *The Origin of Species*. 'The affinities of all the beings of the same class have sometimes been represented by a great tree', he wrote in the chapter on 'Natural Selection'. 'I believe this simile largely speaks the truth. The green and budding twigs may represent existing species; and those produced during each former year may represent the long succession of extinct species. At each period of growth all the growing twigs have tried to branch out on all sides, and to overtop and kill the surrounding twigs and branches, in the same manner as species and groups of species have tried to overmaster other species in the great battle for life.'

is apparent that genetic analogies are not terrifically constraining. Nevertheless, as Dawkins warns, the gene–meme analogy may be taken too seriously (Dawkins 1986: 196), and a good case can be made for avoiding the whole business of neo-Darwinian metaphors, if only because they always seem to muddy the waters. ('Evolution is to analogy,' Steve Jones has suggested, using a memorable analogy, 'as statues are to birdshit'.)

Arguing by analogy certainly has its dangers. There is the risk that a metaphor will come to be treated as if it were a homology. Because the mysterious but fascinating A seems to be rather like the more familiar B, one may be tempted to interrogate B in order to discover the nature of A. And in practice, from time to time, writers simply transfer a checklist of attributes of the gene to the meme. Whatever qualities the gene may have must be mirrored in the meme. The meme is consequently in some danger of becoming a gene in drag. It would then seem to follow that the evolutionary process must work in the same way for memes and for genes. And the obvious conclusion is then drawn, that a science of culture should be modelled on neo-Darwinian biology.

'The main reason we are interested in using the inheritance system analogy is practical', write Boyd and Richerson. 'To the extent that the transmission of culture and the transmission of genes are similar processes, we can borrow the well-developed conceptual categories and formal machinery of Darwinian biology to analyse problems'. (Boyd and Richerson 1985: 31.) It all sounds so pragmatic, so scientific, so reasonable that it is easy to forget that it is all a matter of metaphor and simile. To base methodological conclusions on these loose analogies is reminiscent of what James George Frazer, in *The Golden Bough*, called sympathetic magic. It is like sending black smoke into the sky in order to make rain (if I may be permitted an analogy).

But there is a further, more fundamental difficulty. The actual existence of B may be in doubt, or rather it may come to life only within the metaphor (like the ghost in the machine). Memes are rather shadowy entities, which acquire a certain solidity only by virtue of a metaphorical relationship with genes. (I may not be sure what a meme is, but I think I know what it is like.) Ironically, the gene itself was once thought to be invisible, perhaps only a notional entity. There was a phase in which the gene was granted a distinct material identity, but while DNA and chromosomes are now very much part of the natural world, there are some theorists—Dawkins a notable example—who insist that the gene is a theoretical artefact, a stretch of DNA with the qualities that Dawkins attributes to what he calls a replicator. As he notes, this has got him into hot water with some geneticists, Gunther Stent describing it as a 'heinous terminological sin'.[4] Perhaps Dawkins's notion of the gene will triumph, but one does not have to be an overcautious empiricist to feel uneasy when confronted with a Platonic idea of a thing, which can be grasped only by imagining another idea.

Do cultures evolve?

It may be that the very idea of a science of cultural evolution is misplaced. At the very least, it will all depend on what is meant by culture. And, to be sure, on what one means by evolution. In any case, evolutionary—or Darwinian—approaches to culture, or society, or humanity are not to be reduced to a single question, let alone a single type of answer. The Darwinian programme in the human sciences should be open, eclectic, and multi-faceted.

One of its subjects is the history of the human species, which is what most people mean by human evolution. Another set of questions has to do with the application of evolutionary theory to this history. The relevant aspect of theory is usually taken to be natural selection, although Darwin himself gave equal time to sexual selection in *The Descent of Man*.

If 'culture' in some sense is granted an independent role in this history, then rather different kinds of theory might well come into play. A lot depends on how culture is defined. But one thing that is constant is the notion that culture transcends the individual, that it is a collective property. This introduces a theoretical complication. If culture is collective, and if culture plays a role in human evolution, this seems necessarily to imply a form of group selection. Darwin himself remarked that being a good citizen might have a high cost for the individual, but that good citizenship might be selected because it benefits the community (1871: 203, emphasis added):

It must not be forgotten that although a high standard of morality gives but a slight or no advantage to each individual man and his children over the other men of the same tribe, yet that an increase in the number of well-endowed men and an advancement in the standard of morality will certainly give an

---

[4] See Dawkins's discussion of this issue in *The Extended Phenotype* (1982: 85–9).

immense advantage to one tribe over another. A tribe including many members who, from possessing in a high degree the spirit of patriotism, fidelity, obedience, courage, and sympathy, were always ready to aid one another, and to sacrifice themselves for the common good, would be victorious over most other tribes; *and this would be natural selection.*

If culture is taken to mean the specific traditions of a local community, then another set of 'evolutionary' questions come into focus. These have to do with the interaction between local ecological constraints and particular technological complexes. In the 1960s, this was a central focus of enquiry in American anthropology, and a number of fascinating studies were produced that pointed out, for example, the ecological consequences of rituals and taboos.

And finally, there is the long-running tradition of enquiry into forms of behaviour which we share with other animals. Lorenz took it up, in his way, and Wilson must surely be placed in the same tradition. It can claim to follow the path first trodden by Darwin in the *Expression of the Emotions*. In principle, this research programme should soon be transformed by advances in genetics, but the advent of that transformation does seem constantly to recede into the future.

My position is a simple one. Every one of these Darwinian and neo-Darwinian research programmes seems to me to be well grounded and potentially fruitful. At the same time, this does not exhaust the range of interesting and potentially (or, surely, actually) fruitful approaches that may be brought to bear to interpret and even to explain various episodes of human history, or to answer questions about the nature and limits of human variability. I am in favour of a neo-Darwinian programme in the human sciences (at least so long as it is eclectic and non-exclusive—see Kuper 1994). However, I do not see where memes fit into such a programme.

Indeed I do not believe that memes help us. To begin with, the analogy between memes and genes is fanciful and flawed. Second, if memes are really what we would normally call ideas (and, perhaps, techniques), then it is surely evident that ideas and techniques cannot be treated as isolated, independent traits. (And Darwinians are surely programmed to pay attention to environmental factors.) Third, ideas and innovations are transmitted and transmuted in ways that are very different from the transmission of genes. (Perhaps for this reason, writers on memes sometimes prefer to suggest that they make their way in the world like microbes. Apparently analogies breed analogies. . . .)

We do not need these exercises in sympathetic magic. There are already well-established techniques for the study of cultural diffusion, ideological change, and technical innovation. At the very least, new methods should be tested against the old, to demonstrate that they produce better results. And that is my final objection to the whole memes industry: it has yet to deliver a single original and plausible analysis of any cultural or social process.

Acknowledgements

I am grateful to Alison Shaw for comments on an earlier draft.

References

Boyd, R. and Richerson, P. J. (1985). *Culture and the evolutionary process.* Chicago: University of Chicago Press.
Cavalli-Sforza, L. L. and Feldman, M. (1981). *Cultural transmission and evolution.* Princeton: Princeton University Press.
Darwin, C. (1871). *The descent of man.* London: John Murray.
Dawkins, R. (1982). *The extended phenotype.* Oxford: Oxford University Press.
Dawkins, R. (1986). *The blind watchmaker.* London: Longman.
Dawkins, R. (1989). *The selfish gene* (rev. edn). Oxford: Oxford University Press.
Gould, S. J. (1987). *An urchin in the storm.* London: Penguin.
Kuper, A. (1994). *The chosen primate: Human nature and cultural diversity.* Cambridge, MA: Harvard University Press.
Kuper, A. (1999). *Culture: The anthropologists' account.* Cambridge, MA: Harvard University Press.
Medawar, P. (1982). *Pluto's Republic.* Oxford: Oxford University Press.

CHAPTER 10

A well-disposed social anthropologist's problems with memes

Maurice Bloch

Memes are a wonderful teaching device for the student who wants to learn about human beings *in general*. They serve as a clear and imagination-stimulating concept for the beginner who needs to understand what makes human culture so very different from types of behaviour that are directly genetically driven. Furthermore, talking of 'memes' bypasses the trap of making culture seem transcendental, mysterious and immaterial. The concept of memes thus avoids the Scylla of sociobiology—which fails to take into account the radical specificity of the human mind and what it implies—and the Charybdis of the dualisms of much philosophy and social science—positions that ultimately refuse to accept human knowledge as a natural phenomenon. This is the right epistemological starting point for those who want to engage in the adventure of anthropology.

The final chapter of Dawkins' *The Selfish Gene* on memes is, therefore, an excellent, general, well-written introduction to the subject of culture. But it also attempts to do something which is much rarer and particularly valuable. It presents matters in a way that makes the reader realize that biologists and social scientists are specialists dealing with different parts of what is ultimately a unitary phenomenon. These different kinds of scientists therefore have to have theories that are congruent. Nonetheless, the difficulties they often have in understanding each other are not simply due to separate styles and traditions, but to fundamental features of the different bits of the single totality they are engaged in studying.

There have been many previous attempts at co-operation between natural and social scientists, but they have usually failed because of the crudest misunderstandings of either the nature of the social and the cultural by natural scientists or of the biological and psychological by social scientists. Meme theory deserves a better fate, yet I am afraid the story so far has not been encouraging. Indeed, we have to note how little success the concept of memes has had among social scientists. The great majority of sociocultural anthropologists would not even recognize the word and, when it is explained to them, they are invariably hostile. The reasons are various and include sheer prejudice for anything remotely 'scientific', as well as a suspicion that any 'biologizing' of culture rapidly becomes a legitimization for racism and sexism. (It is easy to disregard this as being a case of ignorant self-righteousness, but the history of the subject shows that such fears are not wholly unfounded.) Some other difficulties, however, are caused by a lack of understanding of the work of anthropologists by memeticists. The aim of this chapter is to show what some of these failures are, in order to show why memes, as they are presented, will not do. My purpose, however, is to further the kind of dialogue initiated, or reinitiated by Dawkins, so that this type of general enterprise, will, at a future date, be more successful.

Memes and the anthropologist's concept of culture

I noted above that, in many ways, Dawkins' work on memes—and that of other writers who have followed him, such as Dennett—is a good, accessible introduction to what is intrinsic in social and cultural anthropology. This fact, however, will not necessarily endear memetics to anthropologists. *At a general level*, Dawkins and Dennett make very similar, if not identical, points to those which anthropologists have always made about human culture. Thus, in the late nineteenth century, Tylor—although an enthusiastic admirer of Darwin and the founder of academic anthropology in Great Britain—stressed how the potential of the evolved human brain meant that the transmission of information between people had become possible in a new way, through symbolic communication, and that this new way meant that human history had a different character to the history of other animals (Tylor

1881, Chs. 1 and 2). Similarly, the concept of 'culture'—usually attributed to the founder of modern American anthropology, Boas, and which
notion became the core of the subject in that country—is, in its fundamental implications, identical to the idea of memes (Stocking 1968;
Kuper 1988, 1999). Again, the classic American anthropologist Kroeber,
at first a pupil of Boas, similarly characterized culture as 'the superorganic', meaning that it reproduces in a way that is independent of the
reproductive system of the carriers (Kroeber 1952).

Memeticists should therefore not be surprised at the exasperated reaction of many anthropologists to the general idea of memes. Biologists
would react in the same way, if, for example, they were told by a soci-
ologist in 1999, ignorant of Darwin and Mendel, that she had made the
following great discovery: that acquired characteristics in animals and
plants were not biologically transmitted to the next generation, but
rather that there were discrete replicating units of molecular material
that were passed on to offsprings. Further, she was going to call these
units of transmission 'closets', by association to the verb 'to close', in
order to stress the oddity of the fact that these units do not merge and
mix into each other in the process of reproduction.

This analogy is a little unfair, but only just. The memeticists could,
with justification, reply that memes have an advantage over the usual
anthropologist's understanding of culture in that talking of memes
stresses the difference with genes, but also reminds us that this
does not mean that we have, for all that, left the natural world behind.
After all, such formulations as Kroeber's 'superorganic', referred to
above, rapidly lead to various forms of implicit dualisms. Such
mystifications have been and, once again these days, are particularly
common in anthropology. It is one of the virtues of the meme idea
that it guards against this temptation in a way which, nonetheless, retains
the core of the culture concept. This is true, but we should not forget
that many anthropologists have made the same point in a variety of
ways, and were able to do this without ever having heard of memes—
for example, Steward (1955), White (1959), Harris (1968), Godelier
(1984), and Levi-Strauss (1962). Furthermore, such an epistemological
stance—although rarer than it used to be—has not been silenced by
such fashions as postmodernism, with its scientific allergies, and appears
in a variety of forms in recent publications: Bloch (1996), Sperber
(1996), Carrithers (1992), and many others. The detractors of anthropology, who want to argue that we are all out-and-out dualists, seem

always to go back to the same old examples of extreme relativism (which they commonly misrepresent) in order to legitimate their contempt for the subject (Pinker 1998; Blackmore 1999). But, as they do this, they ignore the great majority of anthropological work, which they simply do not know or have heard of at second- or third-hand. Of course, it is difficult to keep up with the literature in other disciplines, not to say one's own, but memeticists have freely chosen to explore *exactly* what anthropologists have been studying for more than a century. As a result, they have no excuse in not finding out what the discipline has to offer. To use an analogy once again: a social scientist who, for some reason, chose to write about photosynthesis, would not be justified in pleading lack of time for not acquainting herself with the botanical literature.

The first point to stress is, therefore, that emphasizing the many dramatic implications of the fact that the evolution of the human brain has meant that information can replicate, persist and transform by means other than DNA, is very valuable. The notion of memes does performs this function for a biological audience, perhaps ignorant of anthropology. But this point has already very often been made by anthropologists.

Falling into old traps

Repeating what has already been said in other words can be useful, especially if the point is particularly important. This is the case with some of the discussion of memes. If memes were only a new way of talking about what anthropologists have meant by culture, the lack of acknowledgements would still be annoying to us, but the educational value of the enterprise would remain. If memeticists want to stress the difference between the transmission of information through genes and memes, then they are in step with traditional anthropology. However, it is obvious that this is not all there is to memes. They also want to stress a fundamental *similarity* between memes and genes. The similarity lies in the fact that memes and genes, although made of different substances, both replicate and are therefore subject to the Darwinian algorithm (Dennett 1995). This inclusion of culture and biology within the same framework has a positive aspect which I have stressed above, but the particular similarity emphasized by the memeticist, I argue, is wrong and misleading. Furthermore, it is wrong and misleading in a way that could have easily been avoided if memeticists had been more concerned with anthropology. The problem which anthropologists immediately recognize with memes lies not so much with the very *general* idea, but with a specific aspect of the theory: the notion that culture is ultimately made of distinguishable units which have 'a life of their own'. Only then would it make sense to argue that the development of culture is to be explained in terms of the reproductive success of these units 'from the memes' point of view'.

Bits of culture?

Memetics implies that human culture is made of discrete bits. This is suggested by the analogy with genes. As with genes, finding out exactly what units are involved has proved—even for the most enthusiastic supporter—difficult to define. But, clearly, this analytical isolation has to be somehow demonstrated, even if such a task is seen as a provisional enterprise needing much future refinement. The reason why this *has* to be done with genes is that the very basis of the modern evolutionary synthesis would be incomprehensible without there being distinct genes which can replicate and be selected for independently of each other. To use Dawkins' famous title, it is necessary for genes to have a 'self' to be 'selfish'. For the same reason, if, following Dennett and others, we are to believe that the same evolutionary algorithm governs meme and gene selection, memes have to be something with a defined existence in the world; they cannot remain an arbitrary unit of analysis, created merely to talk conveniently about the world, but with no clear ontology. There is no real doubt about the ontology of genes. Of course, this does not mean that the nature and boundaries of genes are beyond dispute. But it is clear what kind of things they are claimed to be, and scientific advances have made their existence plausible. Again, this does not mean that genes have to be totally independent from each other. We know that genes form clusters and that this clustering affects the selective potential of each gene. But to talk of clusters also implies that we believe in the separate existence of the constituents. Thus, no modern geneticist would seriously maintain that the genome is a totally seamless continuum, which could equally legitimately be divided up in any way that took the fancy of a particular scholar. Now, if the idea of memes is legitimate, the same rule should apply to culture, that whole formed by memes: it too cannot be a continuous entity. The memeticist must believe that there ultimately *are* discrete memes on which natural selection acts, whether these form clusters or not. The memeticist will most probably recognize that different aspects of culture (memes) are linked and that this will affect the selective history of the units. This is what they mean when they talk of memeplexes. But, again such an idea obviously also requires that the units be somehow objectively distinguishable, even though associated into -plexes.

The question is: is this a reasonable way to represent the knowledge of people—in other words their culture? Is it made up of distinguishable bits? As I look at the work of meme enthusiasts, I find a ragbag of proposals for candidate memes, or what one would otherwise call units of human knowledge. At first, some seem convincing as discrete units: catchy tunes, folk tales, the taboo on shaving among Sikhs, Pythagoras' theorem, etc. However, on closer observation, even these more obvious 'units' lose their boundaries. Is it the whole tune or only a part of it which is the meme? The Sikh taboo is meaningless unless it is seen as part of Sikh religion and identity. Pythagoras' theorem is a part of geometry and could be divided into smaller units such as the concept of triangle, angle, equivalence, etc.

Matters become even more difficult when we turn to such much more typical and important phenomena as, for example, a traditional farmer's knowledge about the weather. It is impossible to convincingly demonstrate that this is made up of a number of actually existing, finite number of discrete bits. How many bits would it include? Is the belief that certain types of clouds are an indicator of hail separate from the knowledge that hail damages crops? Memeticists would perhaps then want to speak of 'memeplexes', but they are no more able to establish boundaries around these memeplexes than around the constituent memes. Is the practice of finishing the main rounds of rituals during the rainy season because the ancestors have so ordained and because the harvest can only take place when the crops are dry, is it a part of the memeplex about the weather, or the religion memeplex, or the naive physics memeplex, or the social memeplex? Or is it that all these things link up into one gigantic memeplex? The answer to these questions can only be totally arbitrary. In reality, culture simply does not normally divide up into naturally discernible bits.

The coherence of culture

This fact raises two fundamental issues. The main one, to which I return in the next section, is the ontological status of memes. The other is the question of the coherence, or otherwise, of culture. To this I now turn.

Whether culture is coherent is at the very heart of what has been, for more than a century, a key theoretical controversy—perhaps the most important and difficult source of anthropological polemics. A mass of writing and research has argued over this question and, although anthropologists are far from any agreement, at least we know what kind of arguments need to be taken into account and why the issue is so difficult. This awareness is what seems to be lacking in the discussion on memes, again probably because memeticists have not bothered to acquaint themselves with this work.

A simplified account of the history of anthropology would be as follows. The subject appeared in the academy at the end of the nineteenth century, largely in the wake of the initial enthusiasm caused by Darwin's work. At that time, the discipline saw its role as filling the gap in our knowledge about what happened between the emergence of *Homo sapiens* and the beginning of writing, at which point historians were to take over. The early anthropologists were encouraged by Darwin but hardly Darwinian in any precise sense. In fact, they tended to be guided by a much older tradition which saw the history of humankind as going through a series of 'stages' which had to be passed through in order to reach 'civilization'. Archaeology was to provide information about these earlier times and so too were living non-Western peoples because these were, so it was believed, still 'at an earlier stage'. These stages were usually characterized in a variety of ways, often by their technology. It was thus assumed that, if a contemporary group of people lived by hunting and gathering, studying them would yield information about the early history of mankind, when our ancestors were all hunters and gatherers. This kind of assumption is, of course, still quite common today, not least among sociobiologists, evolutionary psychologists and even memeticists (Blackmore 1999: 195).

However, this type of reasoning soon ran into three very great difficulties. The first is that modern hunters and gatherers live in conditions quite unlike those of our ancestors, precisely because they are surrounded by non-foragers. This means that it is unlikely that what goes for contemporary groups of hunters and gatherers applied also in the past. The second difficulty is that nobody has convincingly shown that such things as religious systems and the technology of food production are closely linked. Thus, modern hunters and gatherers have all kinds of totally different religious systems, and we cannot therefore infer what our ancestors believed in, simply because they were not agriculturalists. Third, the time elapsed since the emergence of *Homo sapiens* is identical for New Guinea highlanders and the people who work on Wall Street; the history of both groups has been equally long, varied, and complex. There is absolutely no reason to believe that the New Guinea highlanders, have somehow been frozen in time and are thus 'living fossils' retaining unchanged the customs of thousands of years ago. We know their history sufficiently well to see that this is simply not so.

These by now familiar difficulties were not, however, the points that the main critics of anthropological evolutionism picked on immediately after the 'evolutionist' period. These early twentieth-century writers chose instead to emphasize that cultural traits *diffused* from person to person, and from society to society. These critics, often enthused by a desire to counter the impious implications of all forms of evolutionism, thus embarked in great enterprises of tracing the geographic itinerary of particularly 'catchy' bits of culture; these they called 'traits' and founded a number of so-called diffusionist schools, such as the *Kultur Kreise* school in Germany, the 'children of the sun' school in Britain, and the American 'culture contact school' to which many disciples of Boas belonged. The principal occupation of these groups was tracing the migration of these cultural traits.

The basic point was quite legitimate. It is that people do not need to go through all the intermediate stages of technological knowledge to be able to use a computer, for example. One generation may have no idea about electricity, while the next may be innovating a new computer program under Windows. This is not due to a speeding up of 'cultural evolution' but the result of a totally different process: the fact that humans can communicate knowledge to each other. In other words, what goes for biological evolution does not apply to culture because humans transmit information from person to person. As noted above, there were many diffusionist schools and several still exist to a certain extent. Some of these schools were somewhat bizzare; others made points that were accurate and interesting. What they have in common, however, is their central argument that human culture is not to be understood as governed by an evolutionary process. Evolutionary anthropologists of the nineteenth century, such as the famous Lewis Henry Morgan who so influenced Marx and Engels, were therefore wrong, because diffusion meant that history was freed from the bounds of nature. It is therefore ironic that the strongly anti-Darwinian flavour of their stance should be so strikingly similar in form to those of the memeticists. It is therefore also particularly relevant to meme theory to take note of the criticisms which the diffusionists soon had to face. These might be called the 'consistency criticisms'.

Culture is consistent

These criticisms came in two forms. The first, American version—associated with such pupils of Boas as Ruth Benedict (1934)—was much influenced by Gestalt psychology. It stressed how cultures form consistent wholes; how every element—wherever it came from—was moulded to fit in with the others because of a psychological need for integration which led to an organically patterned 'world view'. The second type of 'consistency criticism' is more associated with the British school and is usually labelled 'functionalist', although this label itself covers a range of different positions. It settled down into what we may call the 'British social structural approach' which dominated in much of Europe between 1940 and 1970. This approach stressed that culture was not just a set of mental attitudes and beliefs but mental attitudes and beliefs *in practice*: the practice of living in society. And, since living in society implies coordination and ordered cooperation, mental life cannot be separated from the order impressed on it by the nature of society. In this version, the coherence of mental beliefs and attitudes merely reflects the greater and more imperative need to engage in coherent practices necessitated by social structure (Radcliffe-Brown 1952)—and not, as in the American version, due to a psychological need.

Both these approaches inevitably implied a criticism of the diffusionists' emphasis on the transmission of isolated units. The American version of the consistency criticism stressed that even if a bit of information came from one culture and was adopted by another, this could occur if the trait became, in the process, an inseparable part of the culture pattern into which it was incorporated. It then ceased to exist as an identifiable unit. Furthermore, the process of assimilation means that the original element becomes totally modified, so that it was not anymore the same phenomena it had been in another culture. According to this way of seeing things, if one wants to explain the nature of a trait, its ultimate origin is very largely irrelevant. This is, first, because any incorporated trait accepted by an individual or into a new culture was inevitably modified so as to be coherent with the context. Second, any borrowed trait is not a foreign body with a life of its own, but only exists because it is given life by its incorporation into a new whole. Thus, the fact that the habit of making noodles came to Italy from China does not explain why Italians make noodles. An explanation requires why making noodles seemed, and still seems, right to Italians given their beliefs, symbolism, economy, agriculture and perhaps family organization. This is why Italians want, and do, make noodles. What noodles mean to Italians is therefore quite different from what it means for the Chinese.

This type of position was further developed, and to an extent criticized, in Levi-Strauss' version of structuralism. As with the Americans, the need for coherence also originated for him in the human mind. But his view of the patterning process was more rigorous and, above all, more dynamic than in the theory of such writers as Benedict. For Levi-Strauss, coherence came from the psychological necessity for order, made manifest through specific types of structures (such as tree structures and binary oppositions), which then rendered the combination of units possible. For him, structuring is only the first stage in a generative process in which new forms are continually emerging in a similar way to grammar, whose patterning is merely an enabling device for the production of an infinity of utterances.

The Levi-Straussian position is taken a step further in the work of Sperber, who distinguishes sharply between the act of transmission or communication, on the one hand, and the representations in the minds of the producer of the communication and the person who receives it, on the other. For Sperber, unlike Levi-Strauss, these mental representations are integrated and produced by a private mental process which is of a quite different nature to the historical process of continual cultural creation.

What the approaches of Levi-Strauss and Sperber have in common is a relative distancing from the overemphasis on coherent wholes which characterized the earlier ideas of culture in writers such as Benedict. They are thus in accord with other recent tendencies in anthropology that stress the variety of voices in society rather than the (unconvincingly assumed) cultural unison of earlier writers.

These several criticisms and modifications of what we may call the 'Benedict programme' of a coherent, consistent cultural realm are important. But they should not make us forget that anthropologists, such as Sperber, Levi-Strauss and most of their colleagues—as well as myself—accept the fundamental criticisms formulated by the American consistency theorists against the diffusionists: criticisms which apply with equal force against memeticists. Agreement is focused on the fact that the transmission of culture is not a matter of passing on 'bits of culture' as though they were a rugby ball being thrown from player to player. Nothing is passed on; rather, a communication link is established which then requires an act of *re-creation* on the part of the receiver. This means that, even if we grant that what was communicated was a distinct unit at the time of communication, the recreation it stimulates transforms totally this original stimulus and integrates it into a different mental universe so that it loses its identity and specificity. In sum, the culture of an individual, or of a group, is not a collection of bits, traits, or memes, acquired from here and there, any more than a squirrel is a collection of hazelnuts.

The British version of the consistency criticism of diffusionism shares many elements with the American version. These aspects of the theory are, however, not the ones which principally concern me here. However, one aspect is particularly relevant to the subject under examination since it applies equally as a criticism of memetics. British social anthropologists are typically uncomfortable with the very idea of culture. As their name implies, they would rather stress the social aspect of human life than the cultural. Thus, during the period when American anthropologists were developing theories concerning the need for culture to be patterned, the British used the stress on the social to criticize the idea of culture on the basis that it was too decontextualized from the *practice* of ordinary life.

This emphasis on action made the British suspicious of the idea that what is shared among members of the same society is like a vast, consistent encyclopaedia of knowledge incorporating definitions, rules, representations, and classifications. Of course, British social anthropologists did not, like extreme behaviourists, deny that for social and practical action to take place we clearly have to make use of knowledge—a great part of which is learnt by one individual from another. Nor did they deny that this information was then stored in the mind of the receiver so that such knowledge had to submit to psychological laws. But they also wanted to stress that this knowledge is often implicit; that it does not exist in a vacuum. As a result, it is so intimately implicated in action and interaction that it only exists as a part of a whole, only one aspect of which is purely intellectual in character. To represent culture as a collection of bits of information is thus to forget that most of the time it cannot be separated from practices, to which it relates in a number of fundamentally different ways. As a result, for such writers of the British school as Firth (1964) or Barth (1992), knowledge is of many kinds, occurs at many levels and is never independent of a wider practical context. It is therefore better to consider culture not as a set of propositions but as an only partially conscious resource, or perhaps even as a process used in making inferences which inform action—a process which, in any case, occurs at such a speed as to make it necessarily implicit (Bloch 1998, Chs.1, 2, and 3).

Further, this type of 'culture', on which inferences are based, is often quite at odds with explicit beliefs declared by the people studied or by those who study them (anthropologists for instance), especially when these base themselves principally on the declarations and symbolic aspects of the behaviour of those they observe (see Dennett 1987). With such an attitude, British anthropologists see culture as existing on many levels, learnt implicitly or explicitly in a great variety of ways (e.g., Leach 1954; Bloch 1998). It is not a library of propositions or memes. This type of argument is principally intended as a criticism of American cultural anthropology, which (as we saw) was itself a criticism of diffusionism. But clearly it also applies to the simple diffusionist idea that culture is made up of 'bits of information' that spread unproblematically by 'transmission', where transmission is understood as a unitary type of phenomenon. British anthropologists, including myself, would argue that knowledge is extremely complex, of many different kinds, and impossible to locate, as though it were of a single type. It is not only integrated in single minds at different levels of what is commonly understood by the word 'consciousness', but also inseparable from action.

Conclusion

I have dwelt at length on the criticisms which American and British anthropologists have, in the past, directed against the theories of the

memeticists' predecessors: the diffusionists. The reason for such a historical excursion must, by now, be obvious: it is that the arguments
rehearsed against the latter seem equally valid as criticisms of memetics.
As the American critics of the diffusionists showed, memes, like traits,
will continually be integrated and transformed by the receiver of information. They do not spread like a virus but are continually and
completely made and unmade during communication. The process of
their reproduction is not transmission between passive receptors, as is
the case for a computer virus, but active psychological processes occurring in people. That is where life is, not in the bits. Second, as the
British anthropologists stressed, culture, and therefore 'memes'—if such
things existed—would not be made up of a single isolable type of coded
information, which, even for the sake of analysis, could usefully be
understood as separate from social life. Rather, it would consist of a
variety of types of shared knowledge and coordinations which cannot
be understood outside the context of the practice of life; it is something that involves both internal and external constraints and
contextualizations. This variety of phenomena means that transmission
is of many types and is itself part of practice.

Of course, memeticists will want to argue that they are saying more
than the diffusionists ever did and cannot therefore be dismissed in the
same way. They will bring up the originality of thinking of the evolution of culture from 'the memes' point of view'. And, of course, they
are right, because if they had been able to argue that there *were* such
things as memes, this would have been a fascinating new perspective
on human history. The point is, however, that they have not succeded
in arguing convincingly—any more than the diffusionists had before
them when talking of 'traits'—that there are such things *in the world*
as memes. And so, talk of invasion by the 'body snatchers', to use
Dennett's delightful phrase, is an idea as intriguing, as frightening and
as likely as invasion by little green men from Mars. In other words, if
there are no memes, learned discussions about whether their reproductive process is comparable to that of genes and whether their relative
fitness can ever become an explanation of particular cultural configu-
ration, is simply beside the point.

This seems an entirely negative conclusion, but it need not be. As
we saw, the original stimulus supplied by Dawkins set natural scientists
on the path to dealing with the key problems which anthropologists
have been grappling with since the inception of the discipline in the

academy. This was fruitful because, unlike what is the case for most contemporary anthropologists, it made them seek again an *integrated* theory of human evolution which included culture but which did not refuse its special character. This reflection has moved things on from the natural science side, as the work by a number of writers—not necessarily memeticists—shows. It is unfortunate, however, that these people did not make a serious attempt to find out what had been done before concerning these questions; it would have saved them time.

The role of the social and cultural anthropologists in what should have been this joint enterprise is, however, much more shameful than that of their natural science colleagues. They have simply refused to pay attention to people they considered merely as intruders. If they had, they would have disagreed with the memeticists, as I have done here, but they probably would have been saved in the attempt from carrying on in a way which, with time, has become theoretically more and more vague, pretentious and epistemologically untenable. This chapter is an attempt to clear the decks for the very enterprise which Dawkins and Dennett propose. Let us take up the challenge and reflect on what went wrong.

References

Barth, F. (1992). Towards greater naturalism in conceptualising Society. In *Conceptualising society* (ed. A. Kuper) London: Routledge.

Benedict, R. (1938). *Patterns of culture.* London: Routledge & Kegan Paul.

Blackmore, S. (1999). *The meme machine.* Oxford: Oxford University Press.

Bloch, M. (1998). *How we think they think.* Boulder, CO: Westview.

Carrithers, M. (1992). *Why humans have culture.* Oxford: Oxford University Press.

Dawkins, S. (1976). *The selfish gene.* Oxford: Oxford University Press.

Dennett, D. (1987). *The intentional stance.* Cambridge, MA: MIT Press.

Dennett, D. (1995). *Darwin's dangerous idea.* London: Penguin.

Firth, R. (1964). *Essays on social organisation and values.* London: Athlone Press.

Godelier, M. (1984). *L'idéel et le matériel.* Paris: Fayard.

Harris, M. (1968). *The rise of anthropological theory.* New York: Thomas Crowell.

Kuper, A. (1988). *The invention of primitive society.* London: Routledge.

Kuper, A. (1999). *Culture.* Cambridge, MA: Harvard University Press.

Kroeber, A. (1952). *The nature of culture.* Chicago: University of Chicago Press

Leach, E. (1954). *Political systems in highland burma.* London: Bell.

Levi-Strauss, C. (1962). *La pensée sauvage.* Paris: Plon.

Pinker, S. (1998). *How the mind works.* London: Penguin.

Radcliffe-Brown, A. (1952). *Structure and function in primitive society.* London: Cohen & West.

Sperber, D. (1996). *La contagion des idées.* Paris: Odile Jacob.

Steward, J. (1955). *Theory of culture change.* Urbana: Illinois University Press.

Stocking, G. (1968). *Race, culture and evolution: Essays in the history of anthropology.* New York: Free Press.

Tylor, E.B. (1881). *Anthropology: An introduction to the study of man and civilisation.* London: Macmillan.

White, L. (1959). *The evolution of culture.* New York: McGraw-Hill.

CHAPTER 11

Conclusion

Robert Aunger

This book seeks to determine whether the idea of memes might provide the foundation for a progressive line of research on cultural diversification and evolution. In this, the concluding chapter, I do not intend to make the definitive statement concerning the future of memetics. Rather, I attempt to make sense of what has gone before, and to find where there might be grounds for coming down on one side or another of the key issues identified by the authors of the preceding chapters. My comments will be arranged by academic discipline, as it is from these varying perspectives that problems naturally come into view. Indeed, the entire book is arranged—by default—in a similar fashion. It works out (roughly speaking) that meme promoters, put first in this book, tend to be biological in background or inclination, while the more critical voices dominating the later chapters tend to come from psychology and especially the social sciences. I follow the same order in setting out my comments here.

Evolutionary theory

Given its origin in the work of the zoologist and evolutionary theorist Richard Dawkins, the memetics literature has continued to exhibit the strong influence of evolutionary biology. Many of the problems with pursuing this line of research therefore arise from the analogy made between genes as the biological replicator and memes as its cultural equivalent.

Explaining cultural similarity

Just because the *'meme'* meme has been highly successful in popular culture and has even appeared in the *Oxford English Dictionary*, it has not yet been established whether memes are a subject worthy of scientific study. The approval of journalists and British lexicographers merely reflects common usage, and establishes memes as a viable folk psychological concept. However, we cannot be certain that memes—considered as a *scientific* concept—exist.

Why is this? Let us step back a moment and take a look at what distinguishes memetics from alternative theories. Memetics asserts that we can take a 'meme's eye view' with respect to the diffusion of culture. The obvious implication is that there is a previously unnoticed agent participating in social communication—something besides just the sender and receiver that needs to be accounted for. In effect, memetics postulates the existence of an evolutionary agent—a replicator—that evolves in accordance with its own interests (which may be independent from those of either the sender or receiver of messages). Most would identify this agent as the message itself. So a meme must be thought of as a replicator which is active during social communication in such a way that it can influence its own reproduction. The problem is that no one has yet identified bits of information with these qualities.

Why posit the existence of such a thing? Because the fact of cultural similarity needs explaining. Everyone has had the experience of someone else expressing opinions similar to their own or behaving like they do. This suggests there are multiple copies of the information underlying that belief or behavior in the population. But how did this commonality arise? Was the relevant information transmitted to them by others? Or, perhaps similar environments caused commonly held information—information placed in people's heads at birth by genetic inheritance—to be expressed by anyone in that situation. Or perhaps each individual learned the relevant piece of information through earlier experiences with their natural surroundings, without having communicated with anyone or possessing that knowledge innately. In effect, there are three standard explanations for cultural similarity

- transmission (cultural evolution through social learning),
- genes (biological evolution), and
- individual learning (which is convergent evolution through the analogue of mutation from a cultural perspective).

Memetics is associated with the first of these. So what we require in order to prefer the memeticists' explanation of cultural similarity is proof that cultures evolve thanks to the non-genetic inheritance of information. The problem, then, is eliminating the other mechanisms (just outlined) which might underlie the regeneration of cultural traits over time, but which do not involve a cultural replicator—or, indeed, social learning of any kind. How can we discriminate between these alternatives?

The more radical evolutionary psychologists (e.g., Tooby and Cosmides 1992) favor the gene option. They would minimize the role of transmission altogether, emphasizing instead the stimulation of innate mental content by potentially simple ecological stimuli. In essence, they believe 'cultural' traits are already in the brain, with only an environmental spark required for them to be expressed. What remains to be explained from the evolutionary psychological perspective is not social transmission dynamics, but recall dynamics: what kinds of responses do different environments cause to arise? Boyd and Richerson (this volume) disposed of this possibility by arguing that the corpus of human knowledge accumulates too rapidly to be purely genetic in origin. So it seems unlikely that genes—through the instrument of the adapted mind—will single-handedly account for culture. Boyd and Richerson (this volume) also claimed that individual learning in similar environments is an inadequate explanation of cultural similarity, because groups living in the same environment display different suites of cultural traits. This seems, therefore, to leave us with only the cultural transmission explanation—in effect, memes must be invoked to explain cultural similarity. So what is all the fuss about?

In fact, there is another possibility (which is not standard, so I did not include it above): niche construction and ecological inheritance (see Laland and Odling-Smee, this volume). Cultural groups living side-by-side may not be living in the same *effective* environment, because they have modified their natural surroundings in distinct ways. In this case, people learn their cultural traits through interactions with artefacts, rather than other people. Cultural groups living in the same environment differ, in this view, not because they learn the beliefs and values that distinguish them from each other, but because they are influenced by artefacts inherited from previous generations. These can even be types of artefacts which do not communicate information from ancestors to the present-day inhabitants of those environments (like books do). Instead, they could take the form of tools and the 'built' environment, which only *indirectly* influence attitudes and beliefs. So by invoking our ability to manipulate the environment over the long term (an ability we share with many other species), we can continue to discount the role of memes in explaining culture acquisition—even in the face of rapid technological improvements such as surround us today. The feed-forward effects of ecological inheritance, coupled with big, evolved brains able to manipulate the information stored by the activity of previous generations in the environment, can in principle explain the similarity within, and differences between, cultural groups.[1]

Indirect evidence of memes?

In the face of these competing schools of thought, each with vocal and sophisticated adherents, I suggest that for the memetic hypothesis to be favored, we require evidence of some kind that memes exist. This evidence can either be direct or indirect. From indirect evidence, one can infer the existence of memes from the marks their activity leaves behind in the world; direct evidence would show us where memes are and what they look like.

Good indirect evidence for memes would consist of establishing that there is an *independent dynamic* to cultural change which cannot be assigned to the goal-directed activity of people. One would need to observe a directionality to cultural change which reflects the interests of a replicator battling with genes for control over human behavior—*memes*. This is why memes are comonly invoked to explain maladaptive cultural traits, why advocates often gravitate toward examples of memes which seem 'irrational' for individuals (like celibacy), and why memes get equated with viruses, to prompt the implication of induced morbidity in 'hosts'. The problem is that—except for the odd trait here and there—culture is overwhelmingly adaptive for people, allowing our species to dominate our home planet in fairly spectacular fashion. If memes are parasites, they must be symbiotic ones.

So, if memes exist, it is more likely that the course of human evolution should reflect the increasingly *interdependent* interests of genes and

[1] This is not the conclusion Laland and Odling-Smee necessarily intend us to draw when they assert niche construction is important, but their framework *can* switch the burden of explaining technology from cultural to ecological inheritance.

memes. An increasingly effective mutualism between these replicators should result in the human species becoming able to explore new ecological niches, thanks to the additional functionality granted to humans by their relationship to the memetic symbiont. In effect, contemporary humans should be exploiting a broader evolutionary 'design space'— or range of life-ways in which they can thrive—than was possible before memes came along.

The obvious manifestation of synergistic niche expansion through gene—meme cooperation is the rapid increase in technological improvements associated with civilization. Indeed, this is what many would say is the best kind of indirect evidence for the operation of memes—their (unspecified) role in artefact development (e.g., Gabora 1997).[2] Boyd and Richerson's examples (this volume) of incremental improvements in tools such as compasses would seem to prove that a series of artefactual forms can exhibit descent with modification—or the passing of information through a chain of exemplars, forming lineages of information transfer and duplication. After all, such intricate implements show evidence of design, or adaptation to particular functions.

But there are still two kinds of explanations for such obvious design. Does it arise because the best-performing tools are *artificially* selected by people to reflect their own needs? Or alternatively, is their design the 'natural' outcome of independent replicators (memes again) working to achieve a higher probability of replication—primarily by becoming more useful to people? In other words, does the evolution of technology reflect the will of people or the interests of symbiotic memes? It should be apparent that it would be very difficult to tease apart such closely twinned hypotheses.

Nevertheless, some memeticists identify artefacts as memes (e.g., Blackmore 1999; Conte, this volume; Sperber, this volume). Do artefacts fill the bill? As Sperber (this volume) has forcefully argued, three criteria are required for replication—causal efficacy, similarity, and inheritance. Sperber explicitly means to exclude cases of reconstruction from memetics through the criterion of inheritance. Inheritance here means that the information leading to the copy being produced must be

---

[2] Some (e.g., Blackmore 1999) see the development of human language as an early example of gene–meme cooperation: 'digitized' signalling achieves increased fidelity for memes during transmission, while grammatical structure allows increased sophistication in message-passing for human social coordination. But language is subject to the problem associated with any form of communication: the need to transform brain language to a public code and back again. See the section on the communication problem for a discussion of this difficulty.

acquired from the source, rather than having the recipient reconstruct the requisite information for itself. This criterion must hold true whether memes are defined as being in the head or in the form of artefacts.

In fact, there seems to be an impressive array of mechanical replicators out there—chain letters, photocopies, faxes—which meet all of Sperber's criteria for replication. Let us take the case of Web page downloads. Lots of Web pages get visited by Internet 'surfers,' but only in a few instances is the information found there downloaded to the surfer's local hard disk. Presumably, it is some aspect of the content of those pages which triggers the download—and hence the replication of that content. Software ensures that the copying process, based on source information, is high fidelity.

From the 'artefacts' eye view', this is replication, with people placed solely in the role of catalysts for the process. Photocopying is perhaps a more straightforward example: ink-on-paper serves as a template for the copy; there is no phenotypic conversion involved. Indeed, the copying process is just like meiosis: direct replication of the memetic genotype, ink-network to ink-network. Whether anyone reads the copy is immaterial. People are only needed to push the copy button on the machine. (Computer viruses replicate through networked computers with even less human involvement.) The important thing is that there are more copies of the artefact around at the end of the exercise. No replication of information need occur in brains during the process, since each push of the copy button may be produced by some previously installed mental rule about what to find appealing on the piece of paper.

To see artefacts as replicators, we must make a mental flip of perspective, to see the world as the replicator sees it. Dawkins (1976) taught us that we must often think of the biological process from the genetic replicator's point of view (which sometimes means that individual organisms become almost invisible—as in the case of oncogenes, which reproduce themselves through a renegade cell lineage expanding at some cost to the individual's health). So too, from the replicating artefact's point of view, these substances—rather than being external stores of information for the use of people, or aids to getting memes from one brain to another—become the *focus* of a replication story. What is crucial to photocopying, for example, is the original on the glass, the spinning electrostatic drum, and the push of the copy button. The massive human brain is relegated in this story to the trivial task of button-pusher (which a simple-minded automaton could also do). Artefacts *can* in fact inherit information from other artefacts, and 'a scholar is just a library's way of making another library' (Dennett 1995: 346).

Memeticists have invoked mental memes to explain the evolution of technology like photocopiers, but now we have just the reverse: the suggestion that technological replicators are being produced without a necessary role for mental replicators. The 'best' evidence for memes—the evolution of modern technology—turns out to be an instance of replication from another point of view, and hence cannot be used to support the hypothesis of brain-to-brain replication. Rather, technological replication falls into the category of niche construction. Laland and Odling-Smee's ecological inheritance (this volume) may occur through such instances of artefact replication (although it can also occur through the mere survival of existing artefacts, if they last longer than a human generation). And the evolution of the large human brain—a conundrum supposedly explained by memes inciting the construction of a bigger home for themselves, according to Blackmore (1999, this volume)—turns out to be unnecessary for that process to occur, and so must be due to some other cause.

It thus appears that replication is happening all over the place—inside cells (genes), between proteins (prions), and in the environment (artefactual replication). The irony is that it may not be happening in the way originally envisioned by Dawkins—through social learning. Mind-to-mind replication may in fact be the *least* likely mechanism for replication (see the section on memetic phenotypes below). Indeed, whether memes exist in minds remains an open question. Certainly, no model in the memetic literature which makes a brain the site of replication meets Sperber's criteria.[3]

A terminological question now crops up: Should we call this technological replication of patterns on paper or in hard disks a memetic process? Certainly, the information in these patterns does not replicate via imitation, even broadly conceived, and therefore does not fit the original Dawkins/Blackmore definition. If culture is composed of information in people's heads, then the duplication of artefacts does not necessarily help us to explain culture. People may learn from these

---

[3] Actually, there are various models of replication *within* brains (see Delius 1991; Calvin 1996; Aunger 1999), but none that work at the neuroscientific level *between* brains, which is what I'm talking about here.

artefacts or not. Because the present book concerns memes as a contender for the explanation of cultural evolution, I will restrict my use of the word 'meme' to information replicated through social learning (its original context), and leave the question of what to call the technological replication of artefacts to others.

Direct evidence of memes?

So our search for indirect evidence of memes actually led to the discovery of artefactual replicators and grim forebodings about the need for, or existence of, brain-based memes. We come down, I think, to the need for *direct* evidence of memes-in-minds to prefer the meme hypothesis for explaining culture. Since memes are replicators, they must be defined essentially by their means of replication, which should be distinct and independent from those of other replicators (including artefacts). Thus, in my view, the case for memes cannot be made without reference to a mechanism by which information is faithfully replicated through social transmission.

What does 'mechanism of replication' mean in this context? By definition, it is the means by which information exerts some influence over the probability of it being reproduced (Dawkins 1982: 83). One could go further and require a specification of the various resources and their roles in the replication process—the steps leading to the product being assembled, and their speed—but that is no doubt a task for the future.

I therefore conclude that only by finding a mechanism of replication which generates the similarity between people's beliefs and values can instances of inheritance-through-transmission be conclusively distinguished from something like the genetic or developmentalist (evolutionary psychological) alternative. This makes Blackmore's 'existence proof' for memes, as presented in her contribution, unacceptable. It is based simply on the dictionary definition of memes, with a note that this definition implies memes are replicators. In fact, the involvement of memes in the maintenance and diffusion of mental culture remains an open question.

So I submit that meme-promoters will only be proven right about cultural inheritance when someone finds a meme. Nothing except seeing identifiable memes in action is likely to convince people sitting on the other side of the fence that memetics is the best option.

I also think it will be difficult to find a meme without specifying what the search is looking for, and where. Hull (this volume) says we do not need to have a crystal clear definition of memes in order to work with them. He (and Blackmore 1999: 56) cite the oft-mentioned parallel example of genes: that purely operational definitions of genes during the first part of the twentieth century were sufficient for good science to be done. Unlocalized, metaphorical units of inheritance were certainly enough for Darwin to sweep all contenders aside in the nineteenth century, given the logical force of his argument for natural selection as a mechanism. So Hull's admonition to contemporary would-be memeticists is simply to go out and collect evidence of memetic activity in the social world.

Is that going to be good enough? I suggest not. In my view, the situation with respect to cultural inheritance is not the same as that for genes because genes are already established as a mechanism for informational inheritance. Once genes are on the scene, all inheritance, including cultural, might already be accounted for (although I agree with Boyd and Richerson that this is unlikely). If not, then we still have the option of invoking ecological inheritance. So identifying a more-than-operational meme and its mechanism of replication are *both* necessary before memetics can get off the ground. Only by providing a physical model of meme replication can memetics take its rightful position in the list of replicators covered by what Hull terms 'general selection theory.' Until then, they remain simply an analogy to the better-known case of genes.[4]

---

[4] Given this basic level of uncertainty about the nature of memes, it seems to me premature to start making distinctions among memes as a number of authors have done. Plotkin's distinction (in this volume) between 'surface-' and 'deep-level' memes is similar to the standard one in psychology between procedural and descriptive knowledge, or, roughly speaking, between a knowledge of things and how to do things with things. Scott Atran (1998) recently distinguished between 'core' and 'developing' memes. Core memes are acquired through informationally encapsulated modules designed by natural selection; developing memes fall into the cracks between modules, and therefore having to be processed by some amalgam of processing units. Core memes therefore last longer, are more reliably acquired, and generally have the desirable features of good replicators, in Atran's opinion. Such distinctions depend not only on a good knowledge of how encapsulated information-processing algorithms are, but also about how memes might interact with this mental architecture. This makes such propositions doubly 'courageous'. I think we first need to establish that there are memes (as mental entities) before we start dividing them into species (Aunger 1998).

Memetic phenotypes and the communication problem

Even if we ignore these empirical difficulties, major problems remain in meme theory. One is establishing how the genotype/phenotype distinction might work for memes. This distinction is crucial because brains do not directly infect each other with bits of brain stuff; rather, they use signals or messages instead. Brain-to-brain transmission therefore *necessarily* involves the translation of memetic information from brain language to signal language, from one form or code to another, and back again. I will call this the 'communication problem'.

There is also another reason memetics should concern itself with establishing what a memetic phenotype might be. The functional distinction between genotype/phenotype in the genetic system has been generalized by Dawkins and Hull as the replicator/interactor distinction (see my Introduction). Although it is possible for a replicator to also serve as an interactor (as ribosomes do, for example), such a situation is generally considered unlikely to persist. This is because replicators and interactors have fundamentally different roles to play in the evolutionary drama (as store of information and as survivor/transmitter, respectively), and it is usually inefficient for the same entity to play both roles. So a competitor system with independent replicators and interactors would almost certainly win out in an evolutionary race, if only because a more specialized replicator would likely be more robust in its ability to duplicate itself. If memes are considered well-developed replicators, then memeticists will have to develop a notion of a memetic interactor, or 'phemotype' (by analogy to the biological phenotype). While there are a number of contenders for this role, none has achieved widespread recognition.

Part of the problem with developing a rigorous notion of a memetic interactor is coming up with a criterion that positevly identifies it as distinct from its progenitor, the memetic replicator. David Hull (this volume) put forward one criterion for making the distinction between a replicator and its interactor which is generalizable regardless of substrate (and thus a candidate for Universal Darwinism): the relative difficulty of reconstituting the replicator from an interactor. This is a generalization of the Weismannian notion that, in informational terms, you can not go 'backwards' from protein to gene. Such an inability arises because there tends to be some slop in the production of phenotypes: genes do not code for one phenotype, they code for a gradient of possible variant forms (what biologists call a 'reaction norm'), thanks to the impact of environmental conditions on development. So the relationship between replicators and their products is not one-to-one. This implies that information will be lost in the translation from meme to phemotype. It is this loss of information which makes the project of 'reverse engineering' (or inferring the assembly instructions from seeing the product, as Susan Blackmore puts it) so difficult.

Clearing up what is a cultural replicator and what is a cultural inter-actor will also go some way toward avoiding the perennial confusion surrounding 'Lamarckism' in cultural evolution. Since the Lamarckian principle involves the inheritance of *phenotypic* variation, determining whether cultural evolution is Lamarckian depends on distinguishing between memotype and phemotype. Memes may change code or form during transmission, but cultural inheritance will be Lamarckian *only* if the meme is in *phemotypic* (informationally compromised) form during transmission. In this case, the meme-recipient will acquire a phemotypic variant. So making the proper distinction between repli-cator and interactor forms is crucial for basic understanding in memetics.

However, this leaves us in an unfortunate quandary—at least so long as we use information loss as the criterion for identifying the pheno-typic form of a replicator. This quandary arises because, as Hull (this volume) notes, without a clear idea of what memetic information is—that is, how the information in a bit of writing differs from the information in the piece of paper on which it is written—we do not have a good way of determining when it is being lost. If we insist on using information loss as the defining criterion of interactors, progress in memetics will be inhibited until we know how this loss occurs.

Dan Sperber has argued that it is hard for a replicator to solve the problem of information loss during social communication. Artefactual replicators in the form of ink-on-paper can duplicate with very high fidelity: using photocopiers, we have direct replicator–replicator repro-duction, and consequently no loss of information. However, as noted earlier, since bits of brain do not themselves make the journey from one head to another, the memetic life cycle requires that memes be translated from some neuronal construct into another form for social transmission—for example, into parts of speech. Thus, memetic replication cycles involve stages of translation from one code and substrate to another. Since translation is rarely perfect, ths implies that information leakage should regularly occur.

The problem with this is that, if speech is a phemotype, then it is compromised as a message carrier (this is the famous Chomskian 'poverty of the stimulus' argument concerning linguistic message-passing). But then, for the sender's intent to be properly communicated, the receiver must compensate for this information loss by engaging in some kind of reconstitution of the message's intended meaning. However, if there is significant reconstruction of the information content of a meme by each host brain, then the likelihood of message replication is low, thanks to the vagaries of how each brain processes in-coming information (due to the different background information individuals have acquired, the inferencing algorithms they use, etc.).

One way out of this problem, suggested by Sperber, would be for the brain to have a general decoder—a utility enabling it to infer reliably the intention of the sender, and hence the substance of the message, regardless of any intervening noise during transmission or idiosyncracies of sender coding and production. In this view, brains should have evolved filters to assess the utility of information coming in from the social environment to keep us from rapidly being swamped with bad information (or duped into stupid behaviors by people with ulterior motives). This normalizing inferential machine might also ensure the replication of memetic material during social transmission. However, its operation is unlikely to be perfect, so a high mutation rate remains a potential problem.

The need to communicate memes between brains through intermediaries also introduces another, more fundamental, complication. If psychological normalization of memetic inputs is important for successful communication, then memetic information is not, strictly speaking, inherited because it is not passed from person A to person B. Instead, the similarity of socially acquired information between individuals has another cause: inherently structured inferential processing by the brain (Sperber, this volume). These reconstructive processes depend on a long history of genetic selection on the human cortex, not the passing of information from person to person in cultural lines of descent. In effect, the cause of the similarity between the information in A's and B's brains is the result of evolutionary psychology, not memetics. Since the causes are different, one can expect the population-level dynamics to also be different, thanks to differing rates of mutation or types of selection, for example. This creates a fundamental problem for memetics as an inheritance process (the general view of memetics).

However, the memetic process—even if dependent on error correction routines in the brain to produce the cultural similarity of beliefs and values—still confers an evolutionary advantage. This is because the same information is acquired through transmission-plus-correction more efficiently and cost effectively than individual learning through trial-and-error could have done (Dan Dennett, personal communication.). Further, error correction is an important aspect of genetic inheritance as well, so replicator systems can operate with such assistance without having to be called something else.

Susan Blackmore (personal communication) notes that Sperber's reasoning leads to the expectation that, if there is a cultural replicator, there should also be selection for improved mechanisms for its transmission over time. In this way, the reliance on reconstituting information from local resources each iteration would be reduced and the proportion of information actually being transmitted increased. Her presumption is that this is indeed what has happened during the major transitions in cultural evolution, such as language, writing, and computer-based communication. But whether these have increased the transmissability of memes, or merely their copying fidelity, remains to be determined.

Psychology

Another major set of issues concerns the psychology of memes.

Must we go inside?

A fundamental question in this domain is whether memetics can proceed without a clear idea of what kinds of transformations memes might undergo during storage and retrieval by brains. Can memetics leave the brain as a black box, and deal only with social transmission aspects? The virtue of ignoring psychology is that we need not worry about something we do not know too much about anyway: how the brain processes information. This is the line taken by Blackmore (this volume) and Hull (this volume), who argue that memetics can cheerfully ignore what is going on inside people's heads because the real action is happening in the social sphere, or at the level of the population. Boyd and Richerson (this volume), sensing difficulties in this area for memetics—that the psychological mechanisms underlying inheritance are likely to be messy and remain largely unknown—shy away from these particulars. They claim that however the psychological side of things plays out, cultural evolution can nevertheless be seen as a Darwinian process from the level of the population: each generation somehow has to cause information to get stored in the brains of subsequent generations. And it is true that whatever is happening 'inside' can be glossed as some kind of decision-making bias favoring one variant over another during transmission (which is effectively what gene–culture coevolutionary models do). But this does not very effectively limit the kinds of models which need to be investigated.

Further, if memetics disregards psychology, and there are major transformative processes at work in the brain, then memetics is only *explaining* part of the cultural evolutionary process. Since the survival of a meme might depend on an interaction between what happens to it inside and outside the brain, by ignoring one half of the picture memeticists may get the part they explicitly deal with—the public or social part—empirically wrong. Psychologically oriented memeticists generally feel that no social theory, including memetics, can succeed without a proper psychological underpinning.

So, if we agree that we must have a mechanism producing similarity (as I argued above), then we can answer the question of whether memetics must involve itself in psychological issues. The answer is yes: it is crucial that we learn how we learn to become culturally competent members of society. Conte, Sperber, and Plotkin are right, in my view, in this respect. I thus conclude that memetics must peek inside people's heads. Score a point for the psychologists.

Unfortunately, psychologically 'realistic' population-level models of the cultural evolutionary process—whether analytic, or the sort of computer-based simulations preferred by Conte—remain for the future. This is because few social psychologists are interested in filling in the picture with regard to transmission biases, so the wait for greater psychological realism may be a long one.

Imitation

Related to the issue of how memes might replicate is the relationship of memetics to imitation. Two interconnected questions pop up here. First, is a complicated brain essential for imitation? This issue is important because it determines who gets to have memes: only complex intentional agents like people, or more lowly creatures without cortices, such as birds? Many (including Plotkin) argue that there is no consensus concerning the psychological mechanisms of imitation. This is significant because, as Conte (this volume) says, you cannot define imitation without reference to the mental abilities involved. Using behavior as the sole criterion leads to confounds. For example, automatic contagion (such as yawning when others do) is direct phenotypic copying without the inferencing of mental contents. Counting contagion as a kind of imitation suggests that agents do not need to correctly infer another's intention (plus her beliefs and needs, etc.) in order to adopt or imitate her behaviors. What psychological resources imitation demands remains unknown.

The second aspect of the imitation question is a point on which many here voiced an opinion, so it appears to be central. Should memetic transmission be restricted to imitation? Blackmore, citing Dawkins as an authority, restricts memetics to cases of imitative behavior because, she asserts, only imitation serves as a direct copying process, and if memetics is to be founded on replication events, then only imitation can be counted as a memetic mechanism. But as we have just seen, the jury is still out on whether imitation is behavior copying or mental state inferencing (as assumed in the 'theory of mind' literature). This leaves Blackmore's contention somewhat up in the air. Partly on these grounds, Boyd and Richerson, Conte, Hull, Laland and Odling-Smee, and Plotkin make attacks—at least in passing—at Blackmore's position on this issue. It seems that numbers, at least, are against her in this respect. The counter-proposals of Laland and Odling-Smee, Plotkin and Conte are particularly compelling, coming as they do from within the psychological fraternity.

Thus, there is little support for Blackmore's contention that memetics should be limited to imitation because imitation is the only mechanism that can support good replication. It may turn out that directly modeling the behavior of others is not more efficient than independent learning based on environmental cues. Basing meme replication—'by definition'—on imitation, as Blackmore does, is just not going to work.

Imitation is too vague a gloss for what happens during (some kinds of) social transmission. At present, the process appears to involve a magical elision of mental substance from one brain to another—much like the sympathetic transference or 'contagion at a distance' characteristic of 'primitive thought', according to some anthropologists (Hallpike 1979). Once the black box of imitation is opened up, we may find the magic disappears, and rather mundane mechanisms are at work.

Given this general discontent, it seems that any form of social learning, rather than imitation alone, is a better psychological foundation for the cultural evolutionary process. Reader and Laland (1999) take the famous example of milk bottle-top opening by birds as evidence of the need for this generalization. The pecking of bottle-tops has now gone on for many bird generations, and spread through several European countries. Since it is generally felt that birds learn this bit of cleverness not by observing others, but by seeing opened bottle-tops, which inspires their own creativity (a process psychologists call 'stimulus enhancement'), it seemed a pity to exclude such an example from the purview of memetics by limiting it to imitation-based diffusion.

However, if this liberal position on social learning is adopted, many repercussions ensue. For example, the phylogenetic history of memes suddenly becomes considerably longer, with birds and perhaps even more 'primitive' creatures being allowed to have meme-based 'proto-cultures'. In addition, it means that direct contact between hosts is no longer required for memetic transmission, since the source of a meme (such as the tit which pecked a bottle-top) can be absent when a new, naive tit arrives on the doorstep. It is the artefact left behind—that is, the pecked bottle-top itself—which serves as the proximate stimulus for transmission of the pecking meme to the new arrival.

Allowing memes to be learned through any social mechanism implies, then, that memetics must address the issue of artefact production, since memes can be associated with these constructions, and not just brains. Laland and Odling-Smee (this volume) hinted at the importance of artefacts with their concept of niche construction. I discussed their concept from the 'artefacts' eye view' previously. In artefact replication, humans are catalysts—they push the 'start' button which sets the process in motion. But now we see that memes can interact with artefacts as well, in their efforts to find new hosts. I suggest that the involvement of artefacts in a meme's life cycle can be seen as an elaboration of a more primitive process of memetic replication through signaling. In memetic social communication, a human source *produces* the catalyst— a signal such as a gesture or bit of speech—which causes a meme to be replicated in another brain. Such signals are not memes themselves, but rather moving memetic enzymes, produced by memetic activity in the message-sender's brain. On encountering the proper conditions— to wit, an 'innocent' brain—this incoming message instigates the meme replication process in the new host.

This simple model of memetic replication through communication becomes more complex when what might be called a 'communicative artefact' steps into the middle of the communication process. In this case, message-senders create artefacts rather than signals (e.g., written messages rather than speech). These artefacts lie 'dormant' in the environment, during which time they lie in wait for new hosts to infect. For example, words printed on paper can serve as a template for ambient light striking the paper, creating a catalytic signal that passes from the paper-artefact to a naive individual's eyes. This recipient individual, in turn—and in good Sperberian style—reconstructs the meme based on this 'impoverished stimulus', using local mental resources. In this way, memes do not need to pass physically from brain to brain, and so do not need to adopt phemotypic forms such as signals themselves. Nevertheless, the 'replication through communication' problem is solved because a new meme appears in the recipient brain which is causally connected to the source meme through the information provided by the message-catalyst. Sperber's inheritance condition is thus satisfied. And the process is replication as defined by Dawkins (1982; cited earlier) because the message has influenced the likelihood of a meme copy appearing; indeed, that is exactly the role a catalyst should perform in such a chain of events.

But remember that words printed on paper can also be part of a chain of artefact replication, as when they get photocopied. Thus, communicative artefacts are the junction point for two replication processes: that reproducing the artefacts themselves, and that producing new meme copies. So technological development—at least in cases of communicative artefacts—can reflect the evolutionary interests not only of the artefacts themselves, but also those of the memes and people that interact with them. The difficulties of dealing with meme–artefact interactions are rarely discussed in the memetic literature, but such a complex phenomenon obviously requires attention if a comprehensive picture of meme replication is to be achieved.

Mental Darwinism

A second major point of contention is whether the memetic dynamic can be extended into the brain. Can we call individual learning a selection process just like the social transmission process (Changeux 1985/1997)? This proposal has met with considerable disdain, and—at least among academic psychologists—is definitely a minority position (Henry Plotkin, personal communication). Blackmore, in particular, is adamant that whatever is happening inside the head should not be considered part of the memetic process; even if decision-making is in fact selectionist, it still should be treated as an independent replicator system in her view. This may be wise, given the possibility that the meme concept becomes vacuous when extended to cover replication in too many contexts.

However, including selection among alternative mental representations as an intrinsic part of the life cycle of a meme may be crucial to a successful memetics. Two benefits result from this conceptual move. First, only through an analysis of mental properties and processes can good models of transmission mechanisms, such as imitation, be understood. Second, by extending the Darwinian process into the brain ('Mental Darwinism'), one can avoid the confusion of thinking being called 'directed', 'intentionalist', or 'Lamarkian.' Instead of invoking a wholly new kind of process, one can simply suggest that decision-making is selection among memetic alternatives tossed up by some variation-producing process. In the end, the same substrate is involved—neurons. So whether the selection process occurs in the same brain or in different ones, it is all memetic (except that interbrain replication involves the communication problem identified by Sperber above).[5]

Indeed, a great divide separates the Mental Darwinists (usually motivated by evolutionary thinking) from the Intentionalists (usually psychologists or social scientists). Intentionalists do not see any way to avoid issues of meaning when describing human social activity, while the Mental Darwinists argue there is no need to engage in this intentionalist subjectivism to understand memetic processes. Any mental selectionist would prefer to naturalize psychology rather than make the

---

[5] An unfortunate consequence of a selectionist psychology is the appearance it gives of there being no room for human agency in decision-making; that all human psychology is merely a random selection process among alternative behavioral choices. But of course the abandonment of intentionality and free will would be hailed as a victory for memetics by hard-core Mental Darwinists.

many fine distinctions concerning the motivation behind information transmission (such as Conte's elaborate typology in this volume), regarding them as irrelevant to social dynamics.

Sperber agrees with Conte that it is absolutely crucial to distinguish between the commonality of beliefs, values, and emotions that arise through transmission, and those that result from shared individual experiences (such as being in an earthquake) which do not involve any exchange of information between individuals. So the need for causal mechanisms that get information from point A to point B is clear. What remains unclear is whether this necessitates a turn toward intentionalism, or what Dennett (1971) has called the 'intentional stance' (assigning beliefs and values as mental states to others). Perhaps such language is simply a necessary shorthand for discussing psychological processes in big-brained creatures, but should always be understood to be grounded in a Darwinian selection machine at the implementation level.

Social science

To the most pessimistic contributors, both social anthropologists (Kuper and Bloch), memetics is—at best—a promise at present, with no real results to show for itself. The question for these critics is whether memetics will ever contribute anything new to the explanation of society. For a variety of reasons, these anthropologists believe the answer to this question is no.

Ignorance of history

The primary reason for their cynicism is that they believe a quasi-epidemiological approach similar to memetics is already in widespread use in the social sciences, and indeed has a long (but undistinguished) history in those disciplines. In their chapters, both social anthropologists go through an historical account of sociocultural theory in anthropology to argue that memetics is old news—and what is more, bad news. In particular, the idea that some cultures are more stable, or produce a higher quality of life because certain ideas spread better than others, has long been around. Thus, existing explanations for steadfast traditions and similarity in beliefs and values exist which do not invoke memes. However, such evolutionist approaches have been discarded, and superior theories have superceded them. Memeticists miss this 'Big Picture' because they are largely unaware of the comprehensive litera-ture which has accumulated in anthropology concerning cultural change, or the actual history of earlier views such as the cultural diffu-sionists of the early twentieth century (Bloch's target in particular). Being ignorant of the history of diffusionist thought in the social sciences, memeticists are simply condemned to repeat its mistakes.

What remains particularly unclear to these critics is the central claim of memetics: whether there is a novel replicator-based process under-lying the population-level, epidemiological dynamic that is culture change. The primary problem of memetics, from this perspective, is whether there is a new entity on the horizon in whose interests things can be said to happen (the 'memes' eye view'). This replicator would introduce a new kind of functionality which a social institution might serve: that of the memes. As such, it would represent a real and novel alternative to group-level functionalism, or the various flavors of struc-turalist thought current in the social sciences. Unfortunately, the central claim—that a 'memes' eye view' exists—has not yet been proven.

These anthropologists also insist that there is a problem of circu-larity in memetic arguments. Memeticists only study things that seem likely to follow a memetic process, like fashions and fads. The perceived success of such empirical adventures leads memeticists to self-congratulation. But many aspects of culture are not small, isolatable bits of information or practices that readily diffuse in observable time. Take the example of language, which permeates every aspect of culture. How does memetics expect to explain these more fundamental compo-nents of culture?

For some, even the word 'meme' itself provokes problems. Its close parallel to 'gene' may lead memetics astray, they argue, if in fact memes are not the same kind of thing. It also produces a 'revulsion factor' among those who would otherwise be friendly to the Darwinian cause. Memetics is perceived by those outside the brotherhood as an arrogant usurper of territory, making extreme, unwarranted claims. This only serves to put memetics in the same basket with an earlier, related attempt at explaining human social life, sociobiology, which was widely seen as what Dennett (1995) calls 'greedily reductionist'. Sociobiology left no ground for social scientists to stand on, and all the interesting questions were subsumed under a single algorithm: the maximization of biological fitness. This is unpalatable to social scientists not just because of territoriality disputes, but because such reductionism is bound to failure. Thus, an undercurrent in the somewhat scornful reaction of even sympathetic social scientists to memetics is the perception that the social sciences will be 'pre-empted' (Rosenberg 1981) by these evolutionary theories.

But in fact, this threat does not exist, as Plotkin (this volume) is anxious to point out. Can all social processes really be reduced to selection and transmission? The box of concepts available from Darwinism does not impress these social scientists. Memetics seems to employ a very small tool-kit when so many theoretical alternatives are already available and there is so much complexity to explain. In fact, theory abounds in the social sciences. What is lacking is insight into real social processes. Explaining these seems a goal quite far removed from the concerns of most memeticists, who are laboring much further down the organizational hierarchy, worrying about replicators. An uphill battle against a wide variety of other approaches therefore lies ahead for memetics in the social realm.

In sum, memetics is seen as simply another case of those from outside the discipline, in this case largely biologists, 'having a go at explaining culture', but without taking into account many of the complexities this project is widely recognized to entail. The meme critics are happy with the general notion that cultural change involves the diffusion of some vaguely characterized entity, but not with an explanation couched solely in terms of the selection, variation, and inheritance of a particulate replicator.

Confusions about culture

The truly dismaying consequence of this critique is that—as the social scientists themselves admit—they do not have a viable alternative account of cultural change. What memeticists, in their general ignorance of social theory, also do not recognize is that the concept of culture—the very thing which memetics intends to explain—is itself sufficiently problematic that some social scientists advocate its abandonment. In their view, the notion simply covers too complex and varied a set of processes to be useful. (What exactly would replace the concept of culture, or what subconcepts it should be divided into, is not obvious, however.) So, in some sense, the explanatory target memetics aims at—culture—is disappearing into thin air like the Cheshire cat.

At the same time, the anthropological enterprise is itself in serious trouble. So the question suggests itself: Is the major problem with the notion of memes themselves, or with the target it is meant to explain: culture? Those who take culture seriously, as the social scientists do here, find it hard to pin down their own conceptions of this central concept. One can only speak of an impossibly complex tangle of beliefs, behaviors, and social institutions, as well as psychological predispositions and emotions, distributed through all the members of a society. Because all these things are linked, no possibility of reduction is admitted. As a consequence of this conceptual confusion, the project of explaining short-term cultural changes has in large part been abandoned by contemporary anthropologists.

But anthropologists admit that culture is distributed. If we can agree that much of cultural knowledge is socially learned, this implies that such knowledge necessarily diffuses through populations, from individual to individual. All sensory modalities require inputs in the form of temporal streams of information—such as words forming sentences, and sentences paragraphs. At this basic level, individuals therefore must acquire information in bits (which need not be binary). So, something like a unit of transmission must exist. If we can not speak of culture as a phenomenon that can be isolated, then perhaps we can still talk about the problem of how the ideational components of culture are learnt through the social transmission of stimuli. The question then becomes how these units of transmission become translated or incorporated into the body of knowledge and practice that is culture. This is, in fact, Sperber's question—and that of all the psychologists who assert that the psychology of transmission or communication is black-boxed by memetic evolutionists.

So regardless of the complexity of 'culture' as a psychological construct in each person's mind, or as a set of practices and institutions, the informational components underlying the behavioral commonalities of culture (even in the standard anthropological view) must go through channels, migrating from mind to mind. And culture in this form—if you like, exposed to the air as a stream of words, for instance—can be studied in its own right. Indeed, the transmission process—the foundation of how culture, in *all* its manifestations, is maintained—is the rightful domain of memetics. How the bits of cultural knowledge get reassembled once they have reached a new, impressionable mind—another fundamental process—is the rightful domain of psychology. (But, as I have argued earlier, it is also one with important clues about how information acquired culturally is transformed before being sent out again into the social realm.)

The open question is whether these bits of information acquired through social transmission can themselves influence the likelihood they will be further transmitted. Do acquired units of exchanged knowledge have causal efficacy in human affairs independent of the wills of people themselves? In other words, are there memes?

Many arguments in the social sciences still center around the question of 'agency,' or the levels of causation. As Holy and Stuchlik (1983: 2) put it, the question concerning the level at which human behavior is caused is:

basically about the autonomy of agency: if society, or structure, is an objective reality to whose demands people respond in specific ways, then it is an autonomous agency and individual people are its agents, and the ony acceptable explanation is in terms of the functioning of the [social] system. If, on the other hand, society or structure emerges from, and is maintained or changed only by what people do, then individuals are autonomous agents and systems are the consequences of their actions and, in the last instance, explicable by them.

This question—individual or group—has been at the center of the scientific status of social science since its beginnings—with Durkheim, for example, falling on the top-down side of causative directionality, while methodological individualists, such as Rosenberg (1985), fall on the bottom-up side. Dawkins (1976) added a new, 'lower' level of agency to biological theory by emphasizing that adaptations might reflect the interests of genes rather than individuals or groups. Similarly, Dawkins' original meme suggestion indicated that a new, lower level of agency might also be relevant to the explanation of social facts. The 'meme's eye view' shifts the location of cultural agency below the standard 'floors' of individuals or groups to the 'basement' level of information itself. However, such a hoary controversy as that concerning the location of agency is unlikely to be settled here. And even if replication is found to underpin some cultural knowledge acquisition, it is unlikely to be the whole picture, as Sperber (this volume) argues. So it is improbable that memetics will ever provide a full account of cultural change; some aspects of cultural continuity will be due to the push and pull between genes and environment.

There has been an implicit dualism of the agency debate, with the available alternatives have typically been presented as an 'either/or' choice. That is, either individuals are assumed to be fully independent agents, or individuals' cultural repertoires are thought to be fully determined by the society in which they live. A similar restriction has also infected the debate about memes. But in fact it seems likely that individual learning direct from the natural (exogenous) environment can co-occur with social learning, both from other members of the society, as well as from capitalized resources such as books. I think Laland and Odling-Smee's concept of environmental inheritance through niche construction goes a long way toward handling the additional complexity of culture as outlined by Kuper and Bloch. The 'built environment' (including technologies for information storage such as books and computers) which certainly constrains human action, is after all a consequence of the activities of previous generations. Having *three* forms of inheritance (genes, memes, and artefacts) is a means by which a sophisticated theory of mutual constraint relations between individual, societal, and cultural replicator levels can be constructed within an explicitly evolutionary framework.

Progress in memetics?

In the general struggle to understand culture, there is a clear trend for increasing divergence between groups, with decreasing mutual intelligibility. One line is becoming centered around cultural studies, while the other seeks refuge in science. Memes are perhaps more and more likely to be the rallying cry for Darwinists of all stripes when discussing culture, while simultaneously being an object of derision among those inspired by the humanities. Memetics may thus play its small part in the *increasing* division among researchers. Perhaps the debate is not really about memes at all, but rather more a matter of temperament than anything else. At bottom, whether you 'like' memes may be simply due to whether you are a 'splitter' or a 'lumper', a believer in analysis or interpretation.

Although it is obvious that despite the shared belief among those collected here that some kind of evolutionary approach to culture is necessary, significant barriers to communication remain between those from different disciplines. This perhaps derives from the varying histories these disciplines have had with evolutionary approaches. Biologists are predisposed to look at issues of transmission because inheritance is central to their subject, while those trained in the social sciences have been more interested in structure and function—which have traditionally been answered without attention to dynamics, much less the more specific question of transmission. Nevertheless, social anthropology has a long history of evolutionary thought, broadly speaking, which has generally not proven successful. Indeed, a common refrain among social anthropologists seems to be 'been there, done that.' It will be difficult for believers in memes to convince these historically mindful and hence reticent social scientists that this time around things might be different. Similarly, it has proven difficult for the anthropologists to explain exactly what went wrong with previous incarnations of cultural evolutionism, or specifically how the memetic perspective is likely to go wrong itself, even if given a clear run at explaining culture.

But other factors besides academic background also seem to dictate use of the word 'meme' in scientific circles. For example, the teams of Boyd–Richerson and Laland–Odling-Smee both use the same formalism for investigating cultural evolution. But one team rejects while the other accepts the idea of particulate, transmissable units of information as necessary components of the explanation of culture. While Boyd and Richerson may be more enamored of the theoretical possibility of non-particulate inheritance, Laland and Odling-Smee appear to be more impressed by the need for replication to effect transmission. Other rejections of memes are probably idiosyncratic or, perhaps, reflect the continuing confusion surrounding the word 'meme'. Given this multi-layered resistance to memetics, it may be wiser to follow the progress of evolutionary cultural studies more generally, rather than the meme idea *per se*, for a true indication of who is winning the battle to explain culture.

Applying memetics

The question of whether memetics has an empirical future remains open. Among partisans and detractors alike, a major disappointment with the current status of the field is the lack of studies in what might be called 'applied memetics'. Hull (this volume) argues that we should all just go out and 'do it'. However, it is not clear that such an approach will be successful if I am correct about the need to identify the responsible mechanisms underlying cultural inheritance. Instead, I would suggest memetics must first establish how cultural traits maintain themselves in similar forms through generations. Perhaps many mechanisms will be involved, as it is possible there will be as many mechanisms as there are media for social learning.

So we need to develop specific methodologies for conducting memetic studies. There should also be more discussion of existing empirical studies that were not undertaken under the banner of memetics but which could be interpreted as falling within the general purview of this incipient discipline.

It may be that it will not be possible to conduct empirical research in this area for the simple reason that the process being investigated is too complex. From my own experience (Aunger 2000), I would suggest that the prospects for fruitful empirical studies in memetics are daunting. Despite dogged concentration on a highly restricted question (transmission of a limited set of beliefs in a 'simple' oral society), and the application of various multivariate statistical techniques, I have been unable to provide a quantitative estimate of the relative strengths of intra- and intergenerational transmission. On the other hand, a rather more limited transmission science may be possible—and valuable. For example, the exact magnitude of selection coefficients are often unknown in biological studies, but also without much interest. What we really want to know is whether selection is directional rather than neutral, and to identify the selecting agent. The answers to these kinds of questions can get us a long way toward an understanding of the evolution of the system under study and may be possible for a future memetics.

At any rate, as even David Hull (this volume) acknowledges, given the extensive theoretical work already accomplished and the high level of current interest in the subject, something substantial can rightfully be expected of memetics in the relatively near term—either by way of correct, novel predictions derived from the meme hypothesis, or proof that cultural entities with the characteristics of replicators exist. This is because the ultimate test—which would pre-empt theoretical objections—is whether memetics can produce novel empirical work or insightful interpretations of previous results. It has not yet done so, but must do so in the near future. Otherwise, it is likely that memetics will be perceived to be a misguided enterprise. The clock is ticking.

Acknowledgements

Thanks to Gillian Bentley, Rosaria Conte, Liane Gabora, David Hull, India Morrison and Henry Plotkin for feedback on an earlier version of this chapter.

References

Atran, S. (1998). Folk biology and the anthropology of science: Cognitive universals and cultural particulars. *Behavioral and Brain Sciences*, 21: 547–69.

Aunger, R. (2000). The life history of culture learning in a face-to-face society. *Ethos*.

Aunger, R. (1999). Culture vultures. *The Sciences*, 39: 36–42.

Aunger, R. (1998). The 'core meme' meme [Comment on 'Folk biology and the anthropology of science: Cognitive universals and cultural particulars' by Scott Atran]. *Behavioral and Brain Sciences*, 21: 569–70.

Blackmore, S. (1999). *The meme machine*. Oxford: Oxford University Press.

Calvin, W. H. (1996). *The cerebral code: Thinking a thought in the mosaics of the mind*. Cambridge, MA: MIT Press.

Changeux, J-P. (1997). *Neuronal man : The biology of mind*. Princeton: Princeton University Press. [Original work published 1985].

Dawkins, R. (1982). *The extended phenotype*. Oxford: Oxford University Press.

Dawkins, R. (1976). *The selfish gene*. Oxford: Oxford University Press.

Delius, J. (1991). The nature of culture. In *The Tinbergen legacy* (ed. M. S. Dawkins, T. S. Halliday and R. Dawkins), pp. 75–99. London: Chapman & Hall.

Dennett, D. (1995). *Darwin's dangerous idea*. London: Penguin.

Dennett, D. (1971). Intentional systems. *Journal of Philosophy*, 68: 87–106.

Gabora, L. (1997). The origin and evolution of culture and creativity. *Journal of Memetics–Evolutionary Models of Information Transmission*, 1. [http: //www.cpm.mmu.ac.uk/jom-emit/vol1/gabora_l.html]

Hallpike, C. R. (1979). *The foundations of primitive thought*. Oxford: Oxford University Press.

Holy, L. and Stuchlik, M. (1983). *Actions, norms and representations: Foundations of anthropological inquiry*. Cambridge: Cambridge University Press.

Reader, S. M. and Laland, K. N. (1999). Do animals have memes? *Journal of Memetics–Evolutionary Models of Information Transmission* 3. [http: //www.cpm.mmu.ac.uk/jom-emit/1999/vol3/reader_sm&laland_kn.html].

Rosenberg, A. (1985). *Philosophy of social science*. Boulder, CO: Westview.

Rosenberg, A. (1981). *Sociobiology and the preemption of social science.* Baltimore: The Johns Hopkins University Press.

Tooby, J. and Cosmides, L. (1992). The psychological foundations of culture. In *The adapted mind* (ed. J. H. Barkow, L. Cosmides, and J. Tooby), pp. 19–136. Oxford: Oxford University Press.

Index

Location references suffixed with 'n' indicate information contained in footnotes.

Machiavellian Hypothesis 30
magic, faith in 178
Marx, Karl 56
MAS (Multi-Agent Systems) 83, 86,
    89–90, 112
mass suicide 179
mathematical theory of behaviour
    161
Mayr, Ernst 17, 46, 143
meaning, acquisition of 172
Medawar, P. 183
medicine, evolutionary 1
meme definition 5–7, 25–6, 48, 163
    cognitive 113–4
meme fountains 32, 33, 34, 35, 37
meme-gene analogy 3, 184–6
meme-gene interaction 136–7
meme-gene coevolution 29, 34, 38, 39,
    40
Meme Machine, The 5, 11, 16, 34, 164
memeplexes 194
memes
    animal possession of 28
    comparison with genes 192
    developing 213n
    evidence for 208–14
    evolution of 130–4
    high fidelity vs. low fidelity 164n
    mental implementation of 88
    mind-to-mind replication of 211
    niche constructing 133
    psychology of 217–23
    as replicators 8, 11, 12, 26, 40, 206,
        212, 213
    scientific *vs.* folk concept 206
memetic agents 87
    cognitive definition 114
memetic drive 11
memetic driving 31–4, 35
memetic environmental interaction 57
memetic phenotypes 58–9, 214
memetic process, cognitive definition
    114
memetic social communication 221
memetics
    as a science 3–4
    social scientific criticisms 223–5
memetic transmission 38, 57, 97
    brain-to-brain 214, 216, 221
    mechanics of 87–8
    post-industrial society and 133

social behaviour, effect on 86
    spread 98–101
memotypic/phemotypic distinction
    215
Mendel, G. 50, 191
Mendelian genetics 45, 47, 48, 50,
    51
mental Darwinism 222-3
mentalism 5, 6
mental representation of English dialect
    156
metaphors 184–5
Midgley, M. 136
milk bottle-top opening (birds) 130–1,
    220
Miller, G. F. 35
mimetic skill 30
mind modification 107
mind viruses 8, 30
    religion as 35–6
*Mneme, The (Die Mneme als erhaltendes
    Prinzip in Wechsel des organischen
    Geschehens)* 50
molecular biology 47, 48
monomorphic mind 46
morality 186–7
Morgan, Lewis Henry 197
moths, peppered 49
motor skill, revolution in 30
Multi-Agent Systems (MAS) 83, 86,
    89–90, 112
multiple-processes in evolution model
    126, 135
music, human enjoyment of 35

natural language 3
    and cumulative adaptation 17
natural selection 1, 124–5, 143, 187
    cellular level 45
    of information 8
nature-nurture issues 59, 60
nest construction 124
New Guinea highlanders 196
niche constructing memes 133
niche construction 16, 122–7, 128, 207,
    211, 228
    agricultural 137–8
    birds 130–1
    hominid 132–3
normative influence 100
norm recognition 84, 105, 106, 108